(cleaver) 3. *Beta vulgaris* (beetroot) 4. *Agaricus arvensis* (horse mushroom)
(courgette) 8. *Matricaria chamomilla* (chamomile) 9. *Macrocarpon* (sugar snap)

Root to Stem

Alex Laird is a medical herbalist with more than twenty years' experience. Trained in biomedicine and plant pharmacology, she runs the only herbal dermatology clinic based in an NHS hospital, at Whipps Cross. A visiting university lecturer, Alex has published research papers on the use of herbal medicine in the NHS. She runs the charity Living Medicine, which delivers practical workshops to inspire and reskill people in growing and using plants and foods to treat everyday ailments and manage chronic disease.

Root to Stem

*A Seasonal Guide to Natural Recipes
and Remedies for Everyday Life*

ALEX LAIRD

PENGUIN LIFE

AN IMPRINT OF

PENGUIN BOOKS

PENGUIN LIFE

UK | USA | Canada | Ireland | Australia
India | New Zealand | South Africa

Penguin Life is part of the Penguin Random House group of companies
whose addresses can be found at global.penguinrandomhouse.com.

Penguin
Random House
UK

First published 2018
001

Warning

Care should be taken when foraging or using wild ingredients as some plants are poisonous and it is
not always easy to differentiate between toxic and non-toxic species. Do not eat any wild plant until
you are 100% sure of its identification. If you are unsure whether a plant is edible, it is important
that you first check with an expert that the plant is safe to eat, and whether it can be eaten raw or
should be cooked. Some plants should never be eaten raw as they are toxic when raw. With any
wild edible plant, first try a very small portion to ensure that you do not experience any allergic or
negative reactions to the plant.

Disclaimer: the information in this book has been compiled by way of general guidance in
relation to the specific subjects addressed. It is not a substitute and not to be relied on for medical,
healthcare, pharmaceutical or other professional advice on specific circumstances and in specific
locations. Please consult your GP before changing, stopping or starting any medical treatment.
For the safe use of plant medicine, consult a professional medical herbalist at either www.thecpp.uk
or www.nimh.org.uk and see more information at www.herbalist.org.uk. So far as the author is
aware the information given is correct and up to date as at 28 March 2019. Practice, laws and
regulations all change, and the reader should obtain up-to-date professional advice on any
such issues. The author and publishers disclaim, as far as the law allows, any liability arising
directly or indirectly from the use, or misuse, of the information contained in this book.

The moral right of the author and the illustrator has been asserted

Text design by Cooper Street Type Co.
Typeset in 11.75/14 pt Garamond MT Std by Jouve (UK), Milton Keynes
Printed and bound in Great Britain by Clays Ltd, Elcograf S.p.A.

A CIP catalogue record for this book is available from the British Library

ISBN: 978-0-241-37121-3

www.greenpenguin.co.uk

To my teachers and patients from all cultures who have shown me that nature is our ultimate teacher.

Contents

CONTENTS

Introduction

Improving our health is something that every one of us can do. Small, simple changes in how we eat, sleep and live can transform our minds and bodies. By shaping our diet and lifestyle to be more in tune with our body's design, we move closer to the natural world and rediscover the daily and seasonal rhythms that govern all of life.

Root to Stem is a seasonal and holistic approach to health that puts plants, herbs and nature at the heart of how we live and eat. It is a new kind of guide that links individual health to our communities and the planet's health to sustain us all.

The root to stem philosophy – health as a whole
The root to stem philosophy is firstly about the benefits of eating 'whole food' in its natural state. This means eating the edible skin, pith and seeds of vegetables and fruit. Why? Because this is where most of the plant's defence chemicals lie, which we and other animals use as medicine. This is what real food is – in its whole form, it is much more than just fuel, and it comes with many therapeutic compounds that

are vital for supporting our bodily functions. If you want a healthy gut and strong immunity, rethink and relish those fibrous fruit and vegetable skins; they reduce inflammatory hormones and give your gut microbes something to chew on to make many of your body's essential chemicals.

Root to Stem is also about using whole herbs or whole parts of herbs, like the flowering parts, stems and leaves, seeds and roots. And herbs grow everywhere – under our very feet! Supermarkets sell parsley, sage, rosemary and thyme, and together with wild nettles and dandelions, which are abundant and free, these are among the most effective herbs in a kitchen pharmacy. They're not just packed with minerals, vitamins and fibre, but their therapeutic compounds can treat common ailments from allergies to infections. Using plants as medicine is handy, saves money, and used wisely is safe. They're easy to grow and fun to forage. This book will show you how we can use them effectively to treat seasonal health problems and reduce disease.

Eating whole foods is how our ancestors ate for millennia. Using nature's free medicines – light, food and herbs – is the fast track to building our resilience to illness. And by eating as much of the plant as we can, not throwing parts away, we reduce food waste and the damage that we are causing our planet in the process.

Variety is the spice of life

We are made of complex systems and it's the very complexity and variety of plants we eat that enables our bodies to function fully. But since the industrial era, our diet has narrowed dramatically. Much of our disease is now nutrition related: worldwide, poor diet is the second leading cause of death after smoking – and we must change this. We eat too little variety and too many processed carbohydrates, rather than a range of whole foods. Polished white rice, refined wheat and other carb-rich grains are really only semi-food. Refining them down from the original whole grain removes most of the nutrients and fibre that are vital for our health. We, not machines, are designed to be the food processor. It also leaves little for us to digest and so raises our blood sugar too fast, which becomes inflammatory and contributes to chronic conditions such as diabetes, heart disease and cancer. But we can prevent and may even treat these conditions by expanding what and how we eat – and feel better in the process too.

A huge number of plants are edible, maybe as many as 100,000 species, but 75% of our foods come from just twelve species. We should consider fast food and refined carbs as a luxury, a treat, while whole colourful fibrous food sustains us long-term.

Eating root to stem

The beauty of eating whole plants in line with the root to stem approach is that they give us medicine as well as food. They contain the macronutrients of carbohydrate, fibre, protein and fat for our body's structure and energy, along with the micronutrients of minerals and vitamins. They also contain an array of phytonutrients (or phytochemicals – *phyto-* means plant in Greek) that we absorb best through eating. These phytonutrients benefit us in a host of ways. They enable our eyes to function, regulate our hormones, prevent damage to our genes' DNA, help our liver to detoxify, and support our immune responses and heart muscle function among many others.

Plants are unique and keep us alive in other ways too. As the Earth's biggest plants, trees are the main lungs of the planet, giving us oxygen and clean air. They also give us wood, fuel, shade, cool air, carbon storage and shelter for wildlife. For our own survival, we need to protect and encourage their growth.

Why eat seasonally?

Life is all about rhythm. Every day our master body clock drives our bodies on a rough 24-hour cycle known as our circadian rhythm (*circa* = about, *dia* = day) – to rise with the light and sleep in the dark. All of this reminds us that we are part of nature. Our master clock syncs with our many other body clocks, including those in our heart, gut and lungs, to tell them when to wake and digest and when to sleep and repair.

Plants, too, are part of our natural rhythms, marking the annual change in seasons. In the north, our four seasons link us to nature, give meaning and structure to our lives and make us feel part of what is a vast and complex universe. Each season has its own characteristic climate and light. The guide includes how to get the most from the light of each season, as this resets our rhythms to make a big difference to how we feel and act. Taking our cue from nature and plants helps us adapt to the changes in climate and minimize the stress we put on our system.

We can get the most from our food too by eating in line with the seasons – and save on waste in the process. *Root to Stem* lists the cultivated and wild foods that peak in each season and are at their most delicious. There's something satisfying eating food of the moment, and especially if we've grown it ourselves or it's local produce. Eating locally and seasonally is an immediate way to connect with our food and how it's grown. Literally every little action will have a beneficial effect. The food itself is fresher and more nutritious, and we save food miles and resources so it is more affordable for all.

What's good for us is good for the planet

Following a root to stem approach to living is not only good for us but good for the planet too, because we are all interdependent. We are part of one unique and complex ecosystem. Everything we do, no matter how small the action, has an effect elsewhere, though we may not see it. Each living thing, including us, relies on many others in a continual cycle of break-down, transformation and reuse of limited resources for food and energy. By widening the variety of the plants we eat, we contribute to the cycle of life. We promote the biodiversity not only of plant species themselves, but also of the variety of living creatures that thrive on them.

As humans, we are now the dominant species on the Earth and have changed it dramatically, especially over the past 200 years with the mass extinction of wild species. Our use of plastic and its devastating effect on sea life is just one example, as we saw in Sir David Attenborough's *Blue Planet*. Another is the loss of pollinating insects, not only bees, but also butter-flies, wasps, flies, beetles, and other pollinators such as bats and birds. We've cut down so many hedges and wildlife areas to make way for large-scale farming that insects are less able to feed and move. This and pesticide use have reduced their pollination of fruit trees, and so growers must now tackle some of this vast task by hand. A shocking 75% of all flying insects have been lost in Germany over the past twenty-five years, and if numbers continue to decline, this will be a major threat to our food sources.

Simply by planting wildflower seeds in the right places we can increase the population of pollinating insects. Using alternatives to plastic straws, bags and cotton buds can make a big reduction to our polluting waste. The buzzwords 'zero waste' and 'sustainability' are concepts that we can actually apply through our connection to food. It's these easy actions that help to safeguard our food supply for future generations.

Our health also depends crucially on the impact of food production on our environment. This is where meat production is both inefficient and damaging. Animal farming takes up 83% of the world's agricultural land, but delivers only 18% of our calories. In contrast, a plant-based diet cuts the use of land by 76% and halves the greenhouse gases and other pollution that are caused by food production. So there is a real urgency for us to eat more plants and cut down meat-eating. We also waste a scandalous amount of food (about a third of all food in the UK). That's about £15 billion a year alone, not counting the food miles of energy and transport costs. Most of our food waste goes to landfill, giving off greenhouse gases as it rots, which heats the atmosphere and adds to global warming. Throwing food away costs each of us £230 a year. So we can actually save £20 a month, or £80 for a family of four, if we just buy and eat less food, use all its parts, store it better, or even grow some of our own and recycle it as compost.

How to use this book

This pocket guide aims to help you make the most of the seasons for your health. It shows what foods and herbs are the best to eat or grow in each season and how they act as medicine in us. Recipes and remedies detail how you can use them for common ailments and to boost immunity. The guide also offers simple steps to eat and live in sync with natural rhythms, to make a real difference to your health and in a way that supports the planet. These include examples of social activities and annual events that reconnect us to nature and food. Such connections reinforce our natural rhythms and bring us pleasure and health in ways that are less obvious.

This is intended to be a companion for you to dip in and out of as the days go by and the seasons change; it aims to inspire you with the intriguing ways in which our bodies and nature function. As you work your way through the book, these key principles of the root to stem philosophy will help you to maximize the benefits you receive.

How to eat for health and to feel good

There is no one diet for all. Simpler is to apply those basic principles that support our body's design, with its rhythms and functions, to reinforce our health:

- Mostly plants, with edible skin, pith, seeds

- Not too much

- Mostly during the day, ideally within 10 hours

- Many colours, the darker the better

- Some raw

- From a variety of plant families

- A new vegetable or fruit regularly

- Something wild every day

Living lightly and more in rhythm with the seasons is what our bodies want to do naturally if we allow them. And flexibility is built into this. There is no one right way for all, as each of us is unique, with differing make-up, tastes and needs. But we now know a great deal about our design, body clocks and the effects of our food and environment on us. The secret is to take time to notice and follow those rhythms **most of**

the time. We're designed to feel better this way, as a prompt to continue.

As a species, we are beginning to recognize our huge impact on the planet and that our survival rests on conserving its resources, which the rest of nature does so elegantly. Just taking some of the small steps in this guide can transform our environment. Everything we learn shows that we flourish by give and take, embracing diversity and complexity in all aspects of life – in our food, social lives, experiences and nature. It's this that strengthens our responses to difficulty and propels us towards a healthy, satisfying and resilient life.

SPRING

(March – May)

SPRING IS A TIME OF REBIRTH AFTER THE cold, dark days of winter. Shoots that are nutrient-rich are pushing up out of the soil, hungry for new light and the fresh air they need to grow. For nature, it's all about new life, sex, energy, greenery and the promise of the warmth to come over the following months. Synchronizing with the natural world, people start to shift from reflection and planning to doing. With vibrant spring growth, food becomes more varied and abundant, so we can expand our choices to eat a rainbow of succulent young vegetables. In essence, this season gives us a chance to spring-clean both the body and the mind after the sluggishness of winter and to prepare ourselves for the summer ahead.

As the clocks go forward an hour in the UK, people cheer! Remember, 'spring forward, fall back'; it's also a reflection of energy change. While the UK time change marks the start of longer days, other countries see a gentle lengthening of daylight as they move into spring.

A BREATH OF FRESH AIR

With more sunshine comes warmth, the unfurling of leaves and the release of airborne microbes and volatile compounds into the clean, fresh air. Such volatile compounds are what make plants and flowers smell, and these aromatic scents abound with the first spring blossoms. These particles and natural chemicals are good for the environment and good for us. The microbes that feed on the surfaces of leaves swirl into the air alongside plant pollens. Now is the time for therapeutic forest bathing, known in Japan as *shinrin-yoku*. This is walking outside in parks and woodland to take in deep breaths of air, to support lung and immune functions, and protect us against allergies and infection.

THE SUNSHINE VITAMIN

The increased daylight enables us to boost depleted vitamin D stocks to healthy levels. Unlike most vitamins that we have to consume, our skin makes vitamin D itself – the action of the sun's ultraviolet B rays transforms a cholesterol precursor in the skin into vitamin D. This vitamin helps the body to absorb calcium and phosphorus from the gut, thereby building strong bones and supporting brain and nerve function; it is also crucial for activating our immune cells to ward off infection.

Most people living in the US and Europe are low in vitamin D due to lack of sun exposure and overuse of sunscreen. While, of course, we do need to protect our skin against sunburn and cancer, our skin needs vital exposure to sunlight for vitamin D production. The excitement of seeing sunshine and feeling its warmth after the long winter can lead to us overdosing on early springtime sun. Three days a week of moderate exposure (around 20 minutes) over spring and summer should store up enough vitamin D to take you into autumn; though you'll need to top up with vitamin D-rich foods as well.

ARE YOU GETTING ENOUGH VITAMIN D?

Calculate your own safe sun exposure for optimal vitamin D by visiting the Norwegian Institute for Air Research's vitamin D calculator (see link in References and Resources, page 204). You'll need to know the following:

- Day and month
- Your location (choose from a dropdown list or key in your latitude and longitude)
- Skin type
- Time of exposure
- Sky conditions (for instance, cloudless or overcast)
- Ozone layer thickness (if known, or just select medium)
- Surface altitude
- Surface type (sand, concrete, snow, for example)

OPPORTUNITIES OF THE SEASON

Open up to the abundance and variety of young and fresh foods that spring brings, particularly green leaves – it's an especially good time to pick wild plants, as some of them only taste good when young and at their sweetest and most succulent. A good example is ribwort: its tiny, lance-shaped leaves in the middle of a clump have a mushroom–asparagus smell, but only at this early stage.

WHY WILD FOODS ARE BENEFICIAL

Compared with cultivated plants, their wild relatives tend to have more protective phytonutrients. Since they have to fend for themselves, they make chemicals in their roots, leaves, seeds and fruit's skin as a defence against infection, the sun's rays and unwelcome predators, and to attract animals to eat their sugary fruit in order to spread their seed and reproduce. These phytochemicals have antibacterial, antifungal or other protective properties for people, and for animals, as well. By eating these wild plants, we benefit from their ingenuity and complexity, developed over millions of years of adapting to their environment. They give us our natural medicine.

SEASONAL FOODS

As we emerge from winter, young plants – sorrel, cleavers and nettles – spring up with their sour or bitter chemicals. These have detoxing properties, which help remove our bodies' waste build-up after the winter. As these are concentrated sources of nutrients, add just a handful of young shoots or leaves to salads or vegetable dishes.

WHAT'S IN SEASON

PLANT FAMILY	EXAMPLES
Moschatel *Adoxaceae*	Elderflowers
Beet *Amaranthaceae*	Beetroot, spinach, quinoa (leaves and seeds), chard Growing wild – fat hen/ goosefoot, orach, sea beet, sea purslane

Onion *Amaryllidaceae*	Leeks, spring onions, chives Growing wild – wild garlic/ramsons
Carrot *Apiaceae*	Early carrots, celeriac, parsley, dill, coriander, aniseed, caraway, fennel Growing wild – sweet cicely, ground elder, cow parsley/wild chervil, wild fennel **(caution: many wild species are toxic, so double-check identity)**
Asparagus *Asparagaceae*	Asparagus
Daisy *Asteraceae*	Lettuce, chicory, radicchio, Jerusalem artichokes, globe artichokes Growing wild – parsley, dandelion, yarrow, marigold/calendula, chamomile, burdock root, sow-thistle
Borage *Boraginaceae*	Borage **flowers only (do not eat leaves – toxic)**
Cabbage *Brassicaceae*	Purple-sprouting broccoli, kale, spring greens, watercress, horseradish Growing wild – winter/land cress, bitter cress, hedge mustard, shepherd's purse

Hemp *Cannabaceae*	Growing wild – hop shoots
Chickweed *Caryophyllaceae*	Growing wild – chickweed
Cucumber *Cucurbitaceae*	Cucumbers
Bean *Fabaceae*	Broad beans
Gooseberry *Grossulariaceae*	Gooseberries
Mint *Lamiaceae*	Mint, rosemary, winter savory, thyme, lavender, marjoram, oregano, lemon balm
Bay *Lauraceae*	Bay/bay laurel
Willowherb *Onagraceae*	Growing wild – willowherb, evening primrose flowers
Rhubarb *Polygonaceae*	Rhubarb, sorrel Growing wild – sorrel, Japanese knotweed shoots **(check for chemical spray)**
Rose *Rosaceae*	Growing wild – hawthorn, rose, meadowsweet, strawberries
Coffee/bedstraw *Rubiaceae*	Growing wild – cleavers, sweet woodruff, lady's bedstraw

Lime *Tiliaceae*	Growing wild – lime leaf shoots
Nettle *Urticaceae*	Growing wild – nettles, ground ivy, deadnettle, self-heal
Violet *Violaceae*	Sweet violet, heartsease

WHAT TO PICK AND EAT IN SPRING

Learning to forage is one of the most rewarding experiences – just find your local forager to learn how. Many so-called weeds in the garden are edible – so it's very satisfying to stride out into the garden, pick them and take them back into the kitchen, plus you can keep those 'weeds' down in the process. Invasive ground elder is a great example; it's from the carrot family, the *Apiaceae*, and tastes mildly of celery. Like many plants in that family, it has cleansing properties and was reportedly introduced as a vegetable to Britain by the Romans. Pick a few tender shoots for a tasty and nutritious boost to a salad, pesto or soup.

Being able to recognize the many plants that are edible, medicinal and free opens up the world of nature to us in an exciting way. It is liberating to gain skills in identifying plants and their families and to see the extraordinary abundance of nature. You can then have the pleasure of sharing food you've gathered with others. (Be sure to wash plants well before eating.)

Wild garlic

A wood carpeted with wild garlic and its white flowers as far as the eye can see may be one of the wonders of the world – it is certainly one of the most overpowering natural smells, so you may well smell it before you see it! Its Latin name is *Allium ursinum*, but wild garlic's other names are ramsons, bear garlic (or *ail d'ours* in France), wood garlic and buckrams. You can harvest wild garlic – for its flowers and leaves – but must not uproot it on public land. In the UK, the National Trust and Woodland Trust now teach the public how to recognize plants and how to forage sustainably. The only word of caution when gathering these broad-leafed plants is not to confuse them with **lily-of-the-valley, which is toxic to eat** – its leaves look similar, though they're shiny on the back, and the flower looks different (a hooded bell shape), and of course there's no garlic smell to entice you!

WILD GARLIC PESTO

So quick to make and easy to adapt, a tablespoon of pesto can revitalize a vegetable soup or risotto when added just before serving.

Take 3 good handfuls of wild garlic leaves (washed), 2 tablespoons of Parmesan or other cheese, 2 tablespoons of toasted pine nuts (or other nuts), and 2 or more tablespoons of extra virgin olive oil. Add a little lemon juice, pepper and salt, to taste, then blend all the ingredients together.

This pesto keeps for about 4 to 5 days in a jar in the fridge – cover it with a little more olive oil once opened – or you can freeze it. Wild garlic leaves will keep in a sealed bag for 4 to 5 days in the fridge. Makes about 3–4 portions.

VARIATIONS: Add other wild greens to the mix – the tops of young cleavers, nettles, chickweed or ground elder – and use other cold-pressed oils instead of olive oil.

Nettles

Arm yourself with gloves and scissors and you're ready to harvest nettles (*Urtica dioica*). Watch out for the stinging hairs – they like catching bare skin. As well as containing almost three times as much calcium as raw kale, nettles are a source of protein and iron and are rich in vitamin A, vitamin C, magnesium and potassium. Their stinging venom contains at least two of the chemicals your brain uses – serotonin and acetylcholine – as well as histamine and formic acid. Fortunately, blending or cooking neutralizes the venom. You can cook nettles just as you would spinach:

- Add the young shoots to any other vegetable, soup or stew for extra nutrition – their fibre feeds our gut microbes to make energy and brain chemicals among other things.

- Make a nettle tea. Wash and cut up a handful of nettle tops, then pour over 3 cups of boiling water. Infuse for at least 15 minutes, ideally longer, to let the flavour develop before drinking. Once brewed, keep your tea in the fridge and drink it cold over a day, if you prefer.

Cleavers

Cleavers (*Galium aparine*) is the traditional spring cleansing plant. It is a member of the coffee family, *Rubiaceae*. More commonly known as goosegrass, sticky willy or the velcro plant because of its little hooks that stick to clothing when you brush up against it, its protein, minerals, vitamin C, oils and phytonutrients make this spindly climber a classic spring detoxer for the liver and lymphatic system. You will find it on the edges of parks (check the local bylaws; see page 205), in allotments or wherever there are plenty of 'weeds' – otherwise known as 'plants in the wrong place'.

Foraging for such plants – or harvesting them from your garden where they may self-seed – gets you out into the fresh air, saves on food transport (reducing your carbon footprint) and creates a delicious and nourishing dish. For the best-tasting nutrition, pick just young 7.5–10cm stems, as the larger square stalks are too tough to eat. You cannot easily buy the seed, perhaps because the plant is so prolific in the wild, but if you'd like to try growing cleavers at home, collect autumn seeds when they are purple and nurture seedlings for the next spring:

- Either juice a handful of very young shoots with some ginger for a ¼-glass shot to be consumed daily for a month, or infuse a handful of crushed plant in half a litre of water overnight and drink each morning as a spring detoxer.

- Choose young stems to steam-sauté and serve as a vegetable dish.

- Alternatively, add a few young shoots or flowers to a salad or any other vegetable, soup or stew for extra nutrition and detoxing powers – their fibre promotes healthy gut function and feeds our gut microbes to make energy and brain chemicals.

Sorrel

Another cleansing plant, this time from the rhubarb family, is sorrel (*Rumex acetosa*), also known as sour dock from the Anglo-Saxon *sur*, meaning sour, and *el*, meaning all. Find it peeping up through the grass – a delicate green leaf with pointed lobes, it almost melts when heated in butter, and combined with egg makes a deliciously lemony omelette. If you enjoy its flavour, you can plant your own garden version.

Elderflower

Yet another reason to be outdoors at this time of year is elderflowers. These bountiful and free flowers may be foraged towards the end of May; you'll find them in towns and cities as well as in the countryside. It may be surprising, but the flowers of the elder tree (*Sambucus nigra*) actually help hayfever sufferers, since elderflowers have a high concentration of flavonoids – phytonutrients that act on the blood vessels and mast cells to reduce histamine release, and so work to counter allergies. They also look and taste gorgeous. The beautiful creamy-white flowerheads reveal their unique taste when added to liquids. Many producers have used elderflowers' floral taste in jam, tonic, gin, champagne, juices and cordials. At home, you can infuse elderflowers in cream for a fragrant pannacotta, or freeze them in water to make lollies or granita. It's also simple to make your own elderflower cordial.

ELDERFLOWER CORDIAL

A classic taste of spring, the heady scent of elderflower is unique.

In a pan, bring 500g of sugar and 1 litre of water to the boil, stir until the sugar dissolves, then remove from the heat. Add the zest and juice of 2 unwaxed lemons and 1 teaspoon of citric acid, then fully submerge 15 to 20 elderflower heads. Cover with a lid and leave for 24 hours. Strain through a sieve lined with muslin or kitchen paper and pour into sterilized bottles. Label with the date and site of harvest. Store in a cool, dry place; once opened, it keeps for about 3 to 4 months in the fridge, or you can freeze it to make ice cubes to use as needed. Makes about 1.5 litres.

Bitters

Plants with bitter compounds – chicory and radicchio, dandelion leaf and root (also antiviral), older varieties of Brussels sprouts and wild lettuce – are cleansers, and many of them can be added to salads. While some bitter chemicals can warn us of a plant's possible toxicity, the bitters in edible plants stimulate our digestive juices to help us absorb nutrients and reduce bloating. We actually have bitter taste receptors not only in our mouth, but also in our gut and throughout our body; these bitters have been found to activate the immune responses in our airways. Introduce bitter foods gradually, to get used to them.

Onions

Using young leeks, onions and garlic, steam-sauté them with some olive oil. Add a tablespoon or so of water after briefly sautéing, then cover and cook for 5 to 10 minutes until just tender. This fast-food technique, common in the Mediterranean, is perfect for cooking almost every vegetable. It prevents their nutrients from leaching into the cooking water, the oil maximizes absorption of fat-soluble nutrients and the garlic increases the dish's antiseptic and immune-stimulating properties.

Add a handful of herbs and top with feta cheese for a light but vibrant lunch; experiment with other combinations to find your favourite. Adding the herbs – whichever you have handy – is a simple way of boosting the minerals and protective phytonutrients in your diet.

Sweet cicely

Bursting upwards in mid-spring is sweet cicely (*Myrrhis odorata*), with soft, green, feathery leaves and fluffy white flowers, a little like cow parsley. Part of the carrot family, it is sweet and delicious, with a mild aniseed flavour. Use a leaf stalk to sweeten rhubarb and, later, gooseberries, instead of sugar. It spreads all too easily in the garden, so give seedlings to friends and family and share your favourite recipes. Keep picking it to make a sweet and delicious tea that will also act as a digestive to reduce bloating and wind. For this, you can use the stalk, leaf and flower – every part of the plant is edible. Collect the seeds in summer and autumn to dry, and crush them to make a tea or to flavour food or vinegar.

CHALLENGES OF THE SEASON

Flu, colds and sore throats can persist after winter into spring as viruses thrive in the warmer weather. As trees and plants come into flower, they release their pollen, heralding the start of seasonal allergies. Several plants have antihistamine or anti-inflammatory properties. The elder is one such plant: elderflower blossoms in late May, while elderberries appear in the autumn in time for the arrival of flu. Nettle is another. The important thing is to start using them as anti-allergy remedies regularly a month or two before your hayfever or allergic asthma usually begins. Check you are keeping your blood sugar levels balanced too, which helps to counter excess inflammation.

FOOD REMEDIES FOR ALLERGIES

When it comes to resisting allergies, what you want to eat more of are those foods with anti-inflammatory properties:

- Beans and lentils

- Oily fish

- Avocados

- Nuts and seeds

- Vegetables and fruit, including edible skins, seeds and pith (the parts usually discarded in fact help to feed gut microbes that support our immune response)

- Red and black berries and other dark-coloured fruit

And those rich in quercetin:

- Onions

- Broccoli

- Peppers

- Berries

- Grapes

- Herbs (dill, ribwort, greater plantain and eyebright)

Consume less of:

- Sugary foods

- Deep-fried fatty foods

- Alcohol

Anti-hayfever and anti-allergy antidotes

- An anti-hayfever and anti-allergy tea – make tea with a handful each of fresh chopped nettles and elderflowers and 1 tablespoon of thyme; if using dried versions, 1 tablespoon of each is plenty. Pour over 1 litre of boiling water, adding some slices of fresh ginger, for taste. Infuse for at least 20 minutes and drink throughout the day. No need to strain out the herbs, keep them in – it gets better the longer you leave it. Refrigerate once cool; it's particularly delicious when cold.

- A calming eye mask for itchy eyes – make a tea by infusing 1 teaspoon of dried chamomile flowers (or a chamomile teabag) with boiling

water in a mug. When cool,
soak cotton wool pads in
the tea and place gently on
the eyes for 5 minutes,
repeating if needed. Keep
the tea chilled (for 2 days,
maximum); when used straight from the
fridge, the cold also helps reduce any swelling.

- A relaxing tea to help sleep and reduce tension –
  make a tea with lemon balm, lemon verbena,
  chamomile, valerian or Californian or red
  poppies.

- Calming scents for stress – a drop of essential
  oil (chamomile, lavender or frankincense) on a
  tissue can relax you in an instant.

- A gargle for sore throats – crush a garlic clove
  into half a glass of water and add a teaspoon of
  lemon juice and some honey. Gargle with this
  mixture three times a day. It tastes surprisingly
  good! **Tip**: to reduce garlic breath, eat some
  raw apple, lettuce and/or mint leaves.

ASTHMA AND ALLERGIC ASTHMA

More than 5 million people in the UK have asthma,
with symptoms including chest tightness and breathing

difficulties. It can be also be seasonal, triggered by spring allergens. These are proteins (pollen, pet dander) or irritants (dust, smoke, cold air and chemical pollutants) that set off excessive inflammatory or hypersensitive reactions. You can check local pollen counts on the Met Office website and prepare yourself with some of the preventative measures below. Be sure to keep hydrated, as having sufficient water in the body helps keep mucus thin and watery – which assists our lungs in expelling it. In addition, avoid alcohol and sugary foods and those with colourings or other additives, such as sulphites, especially fizzy drinks, as these may trigger a reaction.

- Foods to relax the airways – be sure to include foods from the carrot family: fennel, aniseed, dill, parsley and celery. Start each day with oats, berries and seeds rich in lung-relaxing magnesium, vitamin C, flavonoids and more.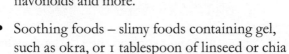

- Soothing foods – slimy foods containing gel, such as okra, or 1 tablespoon of linseed or chia

seeds, soaked overnight in a glass of water;
such foods work by coating and soothing the
gut lining with their gel, and this – by reflex –
relaxes and opens the nearby airways.

- Antiseptic and immune-supporting foods –
garlic, onions, thyme, rosemary – are high in
antiseptic essential oils and reduce the risk of
lung infection through mucus build-up. Along
with mushrooms, they also support immunity
by boosting numbers of the white blood cells
that manage infection.

- Lycopene-rich foods – eat more yellow,
orange and red fruits and vegetables (guavas,
watermelon, tomatoes, papaya, grapefruit,
sweet red peppers, asparagus, red cabbage,
mangos, and carrots), and also asparagus,
daily for their lycopene, a carotenoid,
which stabilizes blood vessels to reduce
inflammatory responses.

- A clearing inhalation – to loosen thick
mucus, inhale a drop or two of expectorant
essential oils (aniseed, fennel, eucalyptus,
rosemary or peppermint) from a tissue.
You can also use electric vaporizers at
home.

- An antiseptic tea – drink a thyme and ginger tea, which thins mucus and aids circulation to clear debris. Pour 1 litre of boiling water over 3 tablespoons of crushed fresh thyme leaves (or 3 teaspoons dried) and some thin slices of fresh ginger. Cover and infuse for at least 30 minutes before drinking throughout the day.

MUSCULAR ACHES AND PAINS

As we get out into the sun with renewed energy for sports and other activities, there's a well-proven kitchen remedy to speed up the healing of muscle and joint aches and pains.

- A DIY muscle cream – make your own capsaicin cream by mixing chilli-infused oil into a base skin cream (or buy capsaicin cream over the counter). The capsaicin is the hot ingredient in chillies. It blocks pain chemicals and opens up the circulation to aid healing. Apply two to three times a day to the skin for quick relief from sprains, strains and tired muscles; it works well on sore backs, too.

SEASONAL GROWING AT HOME

When we grow plants in harmony with their natural life cycle, we can control the quality of the soil and the fertilizer. If these are rich in microbes, minerals and nutrients, this complexity will be reflected in the make-up of the plant, which will benefit us in turn. Moreover, we save precious energy and food miles by harvesting our own food, then composting any waste to fertilize the next crop. All good incentives for us to start growing simple plants that require minimal care – herbs and salad greens, such as rocket, land cress and watercress, which are full of health-giving properties.

GREEN YOUR WINDOWSILL

With the decline in bee numbers making headlines and prompting environmental campaigns worldwide, this is the time to give something back to these precious pollinators upon which our planet relies. By planting herbs such as sage, rosemary, lavender, dandelion, primroses and nettles in a windowbox, you'll be feeding bees, butterflies and hoverflies, without which we could not have a stable and constant supply of food.

AN ETERNAL WATERCRESS BOWL

Watercress – the simplest of all plants to grow. It is a member of the cabbage family and its mustardy phytonutrients (due to its sulphur-containing glucosinolates and indoles) help clear the skin and liver of damaging hormones – make it part of a spring or even year-round detoxing.

- Fill a wide bowl with water and put in a sprig of watercress – you'll notice the little shoots on most supermarket watercress. You can add pebbles for extra minerals.

- Pour in some fresh water every 2 days, watch it multiply, and help yourself to free food!

If you prefer, you can grow watercress in soil, but keep watering regularly – every other day – as it needs to be constantly moist.

GROW YOUR OWN CHILLIES

Chillies are therapeutic in a host of ways, in part through their hot-tasting chemical capsaicin, which warms the body, stimulates blood flow and supports heart function; chilli-based creams lessen joint and nerve pain. Whether you opt for the hair-raising Dorset Naga or the smoky chipotle chilli, every part of these powerful vegetables offers benefits – the seeds are antimicrobial, the pithy membrane (the part richest in capsaicin) has pain-relieving and circulatory properties, and the skin is packed with flavonoid pigments and antioxidant chemicals. Early spring is the time to start growing chillies indoors, as they need warmth.

To grow your own chilli plants from seed:

- Fill a 10cm pot with good-quality seed compost, flatten it down to leave a level surface, then sow a few seeds on top. Most

seeds will germinate, so only sow a few more than you need to account for losses.

- Cover with a fine layer of vermiculite (loose fill chips you can buy at a garden centre), pop in a plant label and water.

- Place a clear freezer bag over the top of the pot and secure with an elastic band.

- After the seeds have germinated, remove the plastic bag and place the pot on a light windowsill. When the seedlings reach 2.5cm tall, move each into its own 13cm pot. Water and place in a light spot indoors.

- When the plants are about 20cm tall (or before if they start to look lean), support them with a stake. Pinch out the tops of the plants when they reach 30cm or so to encourage lots of branches.

- Plant outside in the soil or into a larger pot in late May or when all danger of frost has passed. Water regularly and when the flowers appear, feed every 2 weeks with a general-purpose liquid fertilizer.

PLANTING SNOWDROPS AHEAD

Snowdrops are a much-loved promise of spring, as one of the few brave plants to flower in January. If you order the bulbs now, plant them straight away in shaded, moist soil. Or buy just after flowering when still 'in the green'. But they're best planted fresh from the soil from late April onwards when their foliage is wilting, as they will have a better chance of establishing their growth. Their leaves die down over the summer but their radiance in the gloom of winter is well worth the wait.

BEAUTIFYING YOUR ENVIRONMENT

In springtime the wondrous scents of blossoming cherry and daffodils begin to waft around us, attracting pollinating insects. Bees are back, birds are nest-building and nature is flourishing. And it's a joy to be able to bring some of that natural world indoors.

MOOD-LIFTING FRAGRANCE

A single bunch of daffodils or, better still, the highly fragrant narcissi, divided up into small vases and placed by the bed and around the home, can lift our spirits no end. The volatile oils they release, once inhaled, go straight to the part of our brain called the limbic system (the emo-  tional centre), raising endorphin levels and making us feel good. You can achieve the same effect with oriental lilies or other scented flowers that have multiple stems: put one flower in a small vase in each room of your home to make them go further.

ACTIVELY GOOD FOR YOU

For a free and local (no transport to shops needed) approach, simply picking wild flowers where appropriate and putting them into a vase gives a real sense of delight. The UK plant protection charity Plantlife advises: 'it is not normally an offence to pick the "Four Fs" – fruit, foliage, fungi or flowers – if the plants are growing wild and it is for personal use and not for sale.'

A BEAUTIFUL – AND
HEALTHFUL – FRUIT BOWL

Some fruits and vegetables benefit us more when kept on the worktop, not in the fridge – they not only taste better but also have more health-giving properties. All sorts of produce can gladden the heart when laid out with thought on a plate or in a basket. For instance, red plum tomatoes arranged around lemons on a black plate, or green or red apples in a yellow bowl, make for a stunning display. The skins of apples and tomatoes contain substances called phenols and other antioxidants that increase when stored in the light, but dwindle if refrigerated. Watch as baby plum tomatoes, with their fibre-rich skins, deepen in colour – a sign of higher lycopene levels – and notice their scent as they age. Lycopene is the pigment that gives tomatoes their redness and is a strong anti-cancer and anti-inflammatory compound.

WELL-BEING, MOVEMENT AND SOCIAL CONNECTION

Movement is central to health. We evolved as hunter-gatherers moving about to forage for wild foods and stalk our prey. Yet now we often find ourselves sedentary, stuck behind computers for much of the day. We need physical activity to give our heart muscle and our blood vessels a daily workout to send oxygen-rich blood to our brain and other body organs. Rhythmic movement, such as walking, running, horse-riding, cycling, swimming (wild or in indoor or outdoor pools), dancing and sex, all promote health in various ways and raise our mood. They also help to make a protein that promotes healthy nerve cells called BDNF (brain-derived neurotrophic factor). Good levels of BDNF may improve our mental abilities and is linked to a reduction in depression and anxiety, as well as protecting us from inflammatory diseases like Alzheimer's and Parkinson's.

Spending time with family and friends has been shown to support our health. This social interaction can increase the levels of serotonin and oxytocin in the brain, which both help us feel good and can reduce fear and anxiety.

So, how about combining movement with social connection for even more benefits?

- Mutually beneficial activities – walking outside gets you breathing in deep lungfuls of fresh air, with more oxygen and beneficial airborne bacteria, yeasts and volatile oils (those wonderful scents wafting on the breeze). But why not do the walking while doing something for someone else? The GoodGym, for instance, combines running or walking with a community-based task, such as having a cup of tea with an isolated older person or taking on a task for a community organization.

- Al fresco meetings – don't base meet-ups in a coffee shop, instead stride out into a park or garden. Fresh air leads to fresh thinking – walking and talking can stimulate new ideas and conversation.

- Move faster – even better, take a few minutes to move fast enough to pant, as deeper breathing brings more oxygen to your blood, which then circulates around your body to your brain and other organs. What's more, being sociable raises our natural opiates – the endorphins – and if we smile or laugh, also brings deeper breathing and more oxygen-rich blood to our lungs.

DOING GOOD DOES YOU GOOD

Volunteering has been shown to help us live longer, improve our self-esteem and well-being, and lead to greater life satisfaction. What's more, it reduces levels of depression and stress. The altruistic act of volunteering – actively helping others – is part of what makes volunteering so effective. It also brings a sense of belonging and reduces isolation. The National Council for Voluntary Organisations (NCVO) is a good place to start for ideas and local groups.

A SPRING-CLEAN FOR THE MIND

The best way to clear the mind and focus on the present and the future is to have periods of quiet so that you become aware of what you need. What we call self-care begins with self-awareness for both body and mind, which after all are intimately linked. The following mindfulness exercise is simply a way of slowing down enough to calm our thoughts and gain an inner awareness and ability to learn and grow from our experience. Spend more time if you can each day cultivating this as a new, enjoyable habit.

- Take just 5 minutes each day to sit or lie, breathing naturally, and focus on the sensation of the breath as you breathe in and out. In and out.

- As your mind wanders, which it will do naturally, just bring it back to the sensation of your breath. In and out.

- Notice, as you practise this awareness, how you become calmer and quieter as thoughts and feelings become less insistent.

- Continue to notice your thoughts and feelings, but in a looser way that allows you to identify and address what's needed to move forward.

You can apply this quiet awareness to any activity – walking, cycling, washing up, even cleaning your teeth, for instance.

CELEBRATIONS, ANNIVERSARIES AND SPECIAL DAYS

In the northern hemisphere, the vernal equinox on 19–21 March (depending on the year) marks the beginning of spring in most cultures, when day and night are roughly equal in duration. It tallies with the start of the New Year for Hindus and Iranians (Nowruz). Food plays a key role in these celebrations, and many traditional foods are also therapeutic.

SHROVE TUESDAY OR PANCAKE DAY

Shrove Tuesday is the last day before the Christian period of Lent begins on Ash Wednesday, leading up to Easter; the date varies between February and March depending on when Easter falls). Traditionally Christians would give up rich foods until Easter Sunday. Around the world, it's known as Mardi Gras – fat Tuesday – and celebrated with carnivals. Pancakes were typically eaten on Shrove Tuesday as the last meal using luxury foods, such as eggs and butter, which were then not eaten until Easter.

MOTHER'S DAY OR MOTHERING SUNDAY

Three weeks before Easter, during Lent (so its date varies each year), but marking a break from fasting, daughters would traditionally bake their mothers a Simnel cake – a light fruit cake topped with marzipan – for Mother's Day. You could upgrade a traditional Simnel cake to make it nutritionally richer (substitute ground almonds, hazelnuts or part rye/stoneground flour for some or all of the white flour), but there's another cake that leans on seasonal fruits and herbs that also makes a fabulous gift to bake and share.

RHUBARB ALMOND CAKE

In the rhubarb and almond cake recipe below, sweet cicely offers the sweetness to take away some of the tartness of the rhubarb – both are at their best in spring.

    350g rhubarb
    150g unbleached or brown sugar (or a mix of
        brown and unbleached; or 100g sugar and
        2 large stalks and leaves of sweet cicely)
    3 eggs

220g whole almonds, ground with skins (or mixed nuts, but not Brazils)
10–15 cardamom pods, seeds ground
1 orange (or 1 clementine), zest finely grated
1 teaspoon baking powder

Preheat the oven to 180°C/350°F/gas mark 4 and line a tin with baking parchment.

Chop the rhubarb into 2.5cm lengths and cut the sweet cicely, if using, into a few pieces. Poach very gently with a couple of tablespoons of water and a tablespoon of the sugar, covered, for about 10 minutes until soft. Try to keep a few rhubarb pieces whole – chop these in half and set aside. Discard the sweet cicely, if using, though you can chop it further if you like and include it in the cake with its green flecks. Put the rhubarb into a sieve to drain, pressing out and reserving the delicious pink juice.

In a bowl, beat the eggs with the remaining sugar, almonds, cardamom, orange zest and baking powder.

Fold in the now mushy rhubarb from the sieve, then add the little bits that were set aside – you'll have a pretty liquid mixture. Pour the mixture into the lined tin and bake in the pre-heated oven for 40 to 50 minutes. Place a sheet of greaseproof paper or foil over the surface after about 20 minutes, to prevent burning. Check the cake after 40 minutes with a skewer; it may need longer – when it's done, the skewer should come out clean.

Remove the cake from the oven and, while it is hot, use the skewer to prick holes all over it then pour over some or all of the reserved rhubarb juice. Leave to cool, then turn out. Truly scrumptious served with crème fraîche. Serves about 6.

EASTER

Easter falls on the first Sunday after the first full moon on or after the vernal equinox, and so shifts between March and April across the year. The egg is a symbol of fertility and birth; this healthy protein- and mineral-rich food is also a deliciously portable snack. Traditionally, eggs laid in the Holy Week before Easter

were saved and decorated as Holy Eggs. Painted hard-boiled eggs are still popular in Eastern Europe today, and in the past natural dyes (from onion skins (red and yellow) and leaves (green), for example) would be used to colour the egg shell – plus you can eat them afterwards, so there's no waste.

Chocolate Easter eggs developed in the nineteenth century, but, sadly, many today contain more sugar and fat than cocoa. Yet the cocoa tree (*Theobroma cacao*) itself has medicinal fruits – their flavonoids may help to prevent heart disease and reduce blood pressure.

Chocolate is a fermented food: microbes break down the cocoa beans into many flavour compounds that are easy for us to absorb. The more cocoa chocolate contains, the more bitter its taste, due to cocoa's antioxidant flavonoids. The fermented and roasted bean itself can taste delicious and complex, even before it is shelled and cracked into nibs and processed into chocolate. Small, regular amounts of over 70% cocoa solids in powder or as dark chocolate can be good for you – for instance, studies show that 15g a day of 85% cocoa solids chocolate with a high flavonoid content is therapeutic.

PASSOVER SEDER

This feast – from the 15th to the 21st/22nd day of Nisan in the Hebrew calendar (usually between March and April) – celebrates the liberation of the Israelites from slavery in ancient Egypt and the bountiful offering of new spring vegetables. It features a plate of 'karpas' – vegetables such as parsley, celery or fennel, which are dipped in salt water or vinegar before eating – and 'maror', bitter herbs including lettuce. All such greens are traditionally cleansing by promoting appetite, improving digestion and expelling waste.

HOLI

This two-day colourful Hindu festival of playfulness, mischief, love and forgiveness also celebrates new beginnings; its date is tied to a full moon and often comes in March. Thandai is a spicy, nut milk drink made to celebrate Holi and has many versions using different spices. It's essentially a delicious health-giving drink – its typical fennel seeds, black pepper and coriander are digestive (used to prevent or treat bloating, wind and sluggish indigestion), while rose petals lift the spirits and add a flowery aromatic taste. The coconut, almond or other nut milk or yoghurt are cooling and nourishing. Nuts, pumpkin and sunflower seeds add valuable oils and minerals. Try making your own version by grinding a handful of one or a mix of cashews, almonds and seeds, adding a dessertspoon of one or more of powdered green cardamom seeds, black pepper, fennel, nutmeg, saffron or cinnamon, and a couple of rose petals if you have them. Simmer this gently in any kind of milk (nut, coconut or dairy), then leave to cool. Strain if needed and, if you have no rose petals, add a little rose water at the end to serve.

WORLD EARTH DAY

Occurring every 22 April, this celebratory day started in 1970. The movement now involves 195 countries promoting a healthy environment, with activities that range from making school buildings more green-friendly, to creating new green jobs and investment, and ending plastic, air and water pollution. Perhaps use this day as a prompt to make a pledge to reduce your plastic use or calculate your ecological footprint. Other ways to benefit the planet – while also doing yourself some good – include eating less meat, planting a tree or donating to plant a tree. Check out the tips on its dedicated website to help you 'go green', protect the earth, save money and make every day Earth Day.

SUMMER

(June–August)

AFTER THE PROMISE OF SPRING BUDS, nature flourishes with abundance in summer. Long hot days prompt plants to redirect the sugar energy stored in their roots to flower, leaf and shoot. The full beauty of plants is revealed as their flowers open wide with colour and scents that entice bees, butterflies and other insects to feed on their nectar and pollinate them. Summer is a time of expansion and vitality. Thunderstorms and lightning are most frequent now. In traditional Chinese medicine – a system that sees humans as part of nature – summer is associated with fire, growth, joy and the heart. Now we can embrace the warmth and re-energize.

S ummer's warmth and sunlight improve our physical and mental health. These enable us to function better against diseases such as cancer, heart disease and diabetes. Light triggers a rise in our feel-good chemical serotonin, which boosts our mood. The warmth speeds up our circulation, metabolism and wound healing, and also peps up mental performance. The summer sun maximizes our vitamin D production ahead of the autumn and may help to lower blood pressure, as UVA rays on the skin relax our blood vessels. Outdoor walking is a way to maximize these benefits – to discover more about the beauty and profoundly healing qualities of the natural world.

GETTING THE MOST FROM THE SUN

Now, with more of our skin on show, we want it to look its best. To get glowing skin you need to nourish it from within with food and also outwardly to protect it from over-drying and from sunburn.

Our skin cells form a barrier made from oils and proteins – the oils capture moisture, prevent inflammation and, with the proteins, raise our sunburn threshold (the amount of UV light that leads to sunburn). The following foods nourish skin, keep it looking young and promote rejuvenation; they also have natural sun-blocking action through their oils (or fatty acids), antioxidants and phytonutrients:

- Oily fish (such as sardines, mackerel, herring, pilchards and salmon), rich in the beneficial omega-3 fatty acids, as well as avocados, seaweed, nuts and seeds.

- Extra virgin olive oil, rich in squalene, which is one of the skin's most important protective oils, forming up to 20 per cent of its sebum (oil).

- Skins, seeds and fruits that are black, purple or red (berries and currants), grapes (chew those seeds, as they contain the highest levels), and apples for their powerful proanthocyanidin chemicals, which are antioxidant and help prevent the skin's elastin and collagen, that give skin its shape, from breaking down.

- Green, white and black tea, and cocoa and cinnamon bark, for their tannins and other protective polyphenols.

- Carotene-rich foods (tomatoes, especially when cooked or as tomato purée, carrots, red peppers, squash, sweet potatoes, spinach, kale, apricots and melons).

- Vitamin C- and vitamin E-rich foods; these antioxidants work together to protect the skin's elasticity against light damage. Find vitamin E in seed oils such as cold-pressed sunflower oil and extra virgin olive oil, as well as in almonds and other nuts, wholegrains, dark leafy greens and eggs. Vitamin C is highest in sauerkraut and fermented red cabbage, guavas, all berries (especially blackcurrants), kiwi fruits, watercress and other cabbage family members, and citrus fruits.

- Proteins and peptide-rich foods to support the skin's own proteins, such as collagen, which are natural sun-blockers – milk, eggs, grains, beans, fish and meat.

- Beetroot, turnips, tea, coffee, pea shoots, ribwort – this motley crew all contain the phytonutrient allantoin, which stimulates skin cells to regenerate and heal as well as absorbing harmful UV radiation to lessen the potentially damaging effects to our DNA.

PRACTISE SAFE SUN

With more time spent outside, it's important to protect your skin from sun damage. How much sun is safe, once you've had your daily dose of vitamin D, depends on many factors, including skin type, cloud cover, location, time of day and so on. More protection is needed from midday sun. Notice your shadow – if it is shorter than you (around midday), your exposure to UV is higher; a longer shadow (morning and late afternoon/evening) relates to lower exposure.

After 20 minutes or so of sunshine, a good self-care strategy is 'Slip, Slop, Slap' (slip on a T-shirt, slop on sunscreen, slap on a hat). The British Association of Dermatologists also advises spending time in the shade between 11 a.m. and 3 p.m. when the sun is strong.

Make your own sunscreen

This simple cream includes zinc oxide, a mineral commonly used in sunscreens. The downside of zinc oxide is that it gives a whiteness to the skin. Nevertheless, this cream blocks all UVA and around 95% of UVB with an SPF of 20.

- Blend 20g of non-nano zinc oxide powder (it must be 'non-nano' so that its particles are not small enough to enter the body) with 100g of organic skin cream containing shea butter, almond or other oils.

- Using a hand blender, first combine 20g of aloe vera gel with 20g of non-nano zinc oxide powder, then blend that mixture into 80g of natural skin cream.

OPPORTUNITIES OF THE SEASON

As well as giving long light days to raise our mood and vitamin D levels, summer is the natural time to eat raw foods and salads. The sheer variety of vegetables and fruits available, especially young, juicy roots, makes it easy to eat all the colours of the rainbow.

SEASONAL FOODS

Celebrate nature's bounty with simple, fresh and light meals. Savour young, succulent root vegetables and cooling salads. Leave your desk and eat lunch al fresco – in the park or under a tree – when possible so you get out into the fresh air. In traditional systems of medicine, raw, bitter and sour foods are seen as cooling and offer other valuable medicine. The Chinese recommend eating two-thirds of your food raw in the summer, which helps us manage the hot weather and saves on the heat energy used by cooking.

WHAT'S IN SEASON

PLANT FAMILY	EXAMPLES
Moschatel *Adoxaceae*	Elderberries
Beet *Amaranthaceae*	Beetroot, spinach, quinoa (leaves), chard Growing wild – fat hen/goosefoot, sea beet, orach, sea purslane, marsh samphire
Onion *Amaryllidaceae*	Spring onions, shallots, onions Growing wild/herb – chives
Carrot *Apiaceae*	Fennel, carrots, parsley, dill, coriander, aniseed, caraway, fennel Growing wild sweet cicely, ground elder, cow parsley/wild chervil, wild fennel **(caution: many wild species are toxic, so double-check identity)**
Asparagus *Asparagaceae*	Asparagus

Daisy *Asteraceae*	Lettuce, chicory (endive), radicchio, globe artichokes, cardoons Growing wild – dandelion, burdock root, sow-thistle, sea aster
Cabbage *Brassicaceae*	Kohlrabi, watercress, horseradish leaves, broccoli, cauliflower, radishes, turnips Growing wild/herb – winter/land cress, bitter cress, hedge/garlic mustard, shepherd's purse, sea kale; nasturtiums (*Tropaeoloaceae*)
Chickweed *Caryophyllaceae*	Growing wild-chickweed
Cucumber *Cucurbitaceae*	Cucumbers, courgettes, marrows
Heather *Ericaceae*	Blueberries, wild bilberries, cranberries
Bean *Fabaceae*	Peas, sugar snaps, mangetouts; beans: broad, French, runner
Gooseberry *Grossulariaceae*	Currants (white, black, red), gooseberries

Mint *Lamiaceae*	Mint, rosemary, winter savory, thyme, lavender, marjoram
Bay *Lauraceae*	Bay/bay laurel
Mulberry *Moraceae*	Mulberries
Sweetcorn *Poaceae*	Sweetcorn
Rhubarb *Polygonaceae*	Rhubarb, sorrel Growing wild – Japanese knotweed shoots **(check for chemical spray)**
Rose *Rosaceae*	Strawberries, raspberries, loganberries, blackberries, tayberries, plums, greengages, cherries, apricots, peaches Growing wild – strawberries, raspberries
Lime *Tiliaceae*	Growing wild-lime blossom, lime leaf shoots
Potato *Solanaceae*	Aubergines, tomatoes, potatoes, chillies
Nettle *Urticaceae*	Growing wild – nettles
Violet *Violaceae*	Growing wild – sweet violet, heartsease, pansy

WHAT TO PICK AND EAT IN SUMMER

Almost everything is now growing in abundance, so there's plenty of choice. Parks, gardens, hedgerows and coastal paths offer edible delights to gather while you walk and enjoy en route, or later. Look for chickweed at your feet, hedge or garlic mustard (delicious in pestos or salads) in hedgerows and woodland, the red jewels that are tiny wild strawberries, lemony sorrel leaves, and reach up into the large lime/linden trees to pick their heavily scented blossom for a tea that calms digestion and mind. The young lime leaf shoots are sweet and mucilaginous – a wonderfully descriptive word that means they contain a gel or mucilage that soothes digestive upsets and feeds gut microbes.

On a visit to the coast gather a little marsh samphire, sea beet (beetroot's wild relation), sea kale or sea purslane; take only a few leaves from each plant. The taste of these seaside plants is sure to surprise you – savour their complexity, which comes from growing in a mineral-rich, sea-infused soil. Add a few leaves to a salad for flavour and medicinal benefits. Once home, keep leaves refrigerated in a sealed bag for around 4 days.

Young root vegetables

Carrots, beetroot and radishes are at their most delicate and delicious in summer. And how you choose to prepare them affects their taste, texture and benefits. For example, grated raw beetroot is sweet and juicy (see for yourself in the recipe overleaf), while raw chunks are dry and woody. Grating breaks down the plant cells, liberating enzymes and sugars to make new flavours. What's more, its immune-stimulating polyphenols go up as the plant responds to its 'damage' by releasing protective antioxidant chemicals – the more of those we ingest, the better the benefits for us

Carrots – grated raw, or steamed or boiled until just cooked, then dressed with roasted cumin seeds, olive, sesame or other cold-pressed oil, sherry or other vinegar (or lemon juice) and sprinkled with the edible green tops, make a sweet-tasting and pretty dish. Crumble over feta or goat's cheese to add protein and salt to the carrot's sweetness; the tops can also be used in salads and as a cooked vegetable.

BEETROOT, WATERCRESS
AND FETA SALAD

This sweet and juicy salad gets your taste buds tingling with a tangy dressing. It's also a feast for the eyes and ready in a matter of minutes.

Grate a scrubbed raw beetroot, preferably with its skin still on, unless it's particularly tough, into a bowl, scatter over some feta cheese, a generous handful of watercress (or other mustardy green), including the stems, some torn-up radicchio or chicory, a chopped spring onion and a handful of chopped parsley or other herb. Make a dressing with 1–2 tablespoons of extra virgin olive or other cold-pressed oil, a squeeze of lemon or lime juice or a little balsamic vinegar, salt and pepper. Serves 1–2.

VARIATIONS: You can vary this salad infinitely with other gently cooked or raw vegetables, along with some lightly toasted sesame seeds or other seeds and nuts. If you have any mineral-rich wild coastal leaves, their slightly salty flavour contrasts beautifully with the sweet beetroot. (In the autumn, switch to grating squashes and pumpkins with their skins – the pigments that give them that beautiful orange colour are where the medicine is.)

TIP: The trick to a great salad is a dressing that makes it truly juicy and piquant, so be sure to add enough oil and acid.

Blackberries

Blackberry picking is fun and free. Their black pigment contains anthocyanins, which boost our immunity and protect us against allergies, and the fruit itself is very high in vitamin C. You'll find blackberry hedges in towns and parks as well as in hedgerows in the countryside. So collect a tubful and use them to whip up a cooling and nutritious breakfast for a hot summer's morning.

- In a bowl, mix together 40g of medium or pinhead oatmeal or jumbo oats, a handful of mixed nuts, a handful of mixed seeds including linseeds, 125ml of water or a milk (dairy, nut, coconut, oat, etc.), and 1–2 tablespoons of yoghurt or kefir. Put into the fridge overnight to soften, or leave out to ferment in the warmer temperature and benefit from the free microbes that multiply. In the morning, give the creamy mix a stir and add a handful of blackberries, or mixed berries, along with some extra fruit including their edible skins and seeds, such as apples or pears.

If you've harvested plenty, freeze some blackberries now to use in the winter to boost immunity.

BLACKBERRY DESSERT

Even if bought frozen, blackberries are packed with vitamin C, black flavonoid pigments and fibre, all of which support the skin, gut microbes and immunity. The simplest way to eat them is unadorned (or in the breakfast just mentioned), but why not try this speedy dessert instead?

Arrange the berries in a shallow dish and scatter over some chopped mint. Cover with a layer of crème fraîche (or oat cream) and sprinkle with brown sugar. Pop under a

preheated grill for about 5 minutes, until the
sugar is bubbling.

VARIATIONS: You can use any berries or soft
fruit for a change of flavour, depending on what
other seasonal ingredients you have in the fridge
or fruit bowl.

Summer vegetables
Most summer vegetables are best when prepared
simply with one or two other ingredients:

- Courgettes – grate courgettes, including their
  stalks, raw into salad with a generous dressing
  of 3 parts cold-pressed olive or other oil
  to 1 part cider vinegar (or lemon juice or
  balsamic vinegar), some grain mustard, salt
  and pepper. Mix well until it all emulsifies.
  Alternatively, sauté them very briefly, no
  longer than 3 minutes, in extra virgin olive
  oil (uncovered), adding salt and pepper and
  then some chopped fresh herbs; if you are
  lucky enough to have their beautiful flowers,
  explore recipes for stuffing, then frying or
  baking them.

- Fennel – slice raw into salads, or slice a bulb thickly lengthways, keeping the green fronds separate, then cover and slowly stew in olive oil and/or butter with a slice of lemon and a bay leaf, salt and pepper for about 40 minutes, until soft; sprinkle with the chopped, green fronds.

- New potatoes – boil with mint, or another herb, and serve with steamed runner beans, sorrel, marsh samphire, spring onions or watercress.

- Peppers – other than cutting them up raw in salads, one of the most delicious ways to eat them is to stew them slowly in oil with sliced garlic, a bay leaf, salt and pepper for at least 30 minutes, until soft; include the seeds – they taste fine, contain useful medicine and make the dish quicker to prepare!

- Tomatoes – eat raw or roast them: put ripe, sliced or whole baby plum tomatoes into a shallow baking dish, tuck a bay leaf underneath, then add a slug of olive oil, some crushed garlic, fresh chopped basil, thyme or other herbs, salt and pepper. Grill for 10 minutes, turning the tomatoes if needed, until their juices bubble and run, or roast in the oven for 10 to 15 minutes. Crumble over a salty cheese or add a little chilli or chilli sauce before cooking if you like.

- Globe artichokes – fun to eat with other people; just boil in plenty of water for 30 to 40 minutes until you can pull out one of the leaves easily, then serve with butter and/or cold-pressed olive oil.

QUINOA WITH SUMMER VEGETABLES

Quinoa seeds can be treated as a grain but they are in fact part of the beet family. Quinoa is highly nutritious, gluten free and contains complete protein, which is vital for vegetarians and vegans, about double that of brown rice. The mangetouts and sugar snaps with their pods add more microbe-feeding fibre and skin phytochemicals.

Cover and gently simmer half a cup of red or black quinoa (highest in therapeutic polyphenols) in a pan of water with a couple of whole garlic cloves for 10 to 15 minutes, until cooked and all the water is absorbed. Meanwhile, cut the woody ends off some asparagus and sauté the stems gently in a wide shallow pan with olive oil, a chopped leek, and some mangetouts or sugar snaps. (Don't discard the woody ends of the asparagus; use them for flavouring a stock or making a tea to counter urinary tract infections and joint pain.)

Serve with new carrots, simmered, covered, for about 7 to 10 minutes in a little water, then drained and sautéd in butter or oil and sprinkled with toasted cumin seeds and chopped coriander leaves. Serves 2–3.

VARIATIONS: Broad beans – fresh or frozen – are another easy addition, full of fibre, energy and hormone-balancing nutrients, but do experiment with other summer vegetables.

Summer bitters

Among the most medicinal foods are those with bitter flavours (chicory, coffee and mustard greens) and sour tastes (citrus fruits, vinegar or yoghurt, for instance). Best-known 'bitters' are the plant alkaloids which also provide common drugs such as quinine, once used for malaria and still added to tonic water, morphine for pain, and caffeine, naturally present in coffee. We need some of these bitter compounds from foods as medicine. Our taste receptors are not only on our tongue, but further down the gut, in our airways and throughout the body. Bitter compounds lock on to these receptors, then stimulate our digestive juices and strengthen our immune defences. The bitter herbs gentian and wormwood are used precisely for this reason. As we age, we tend to seek out bitter flavours to boost declining digestive (and other) functions.

Bitter-tasting plants such as chicory have another role, beyond stimulating digestion, liver and immunity, as they are traditionally seen as cooling foods. Raw foods and salads are also considered cooling. Examples of other bitter and cooling foods are:

- Cabbage family

- Daisy family – chicory, radicchio, dandelion, sow thistle

- Coffee

- 'Bitter aloe' refers to the latex lining the *Aloe vera* leaf, so eating a short section (2.5cm or so)

of a whole leaf (or adding it to a smoothie)
helps to treat constipation or bloating as well
as being cooling

- Members of the cucumber family – cucumber,
  and the bitter gourd/karela (*Momordica
  charantia*), known for its anti-diabetic action

- Mint, especially peppermint

- Medicinal herbs – gentian (*Gentiana lutea*) and
  wormwood (*Artemisia absinthium*), used in
  tinctures in drop doses to stimulate appetite
  and reduce wind and bloating

Good health depends on eating a wide variety of
foods to provide a full range of phytonutrients and
enzymes. So the more diverse flavours you eat, including bitter and sour, the better. If you have children,
you can establish healthy preferences for life by giving
them citrus and other sour fruits and bitter salads,
even in small amounts to start with, so they learn to
eat across the spectrum. If you find bitters a little
harsh, you can mask them with some honey or put
them into a smoothie with a prune or date, then over
time reduce the sweetness until you're a convert.

Many cultures take full advantage of bitter compounds and all that comes with them to stimulate
digestive juices and strengthen immune response.

The Persians have *sabzi*, an enticing plate of crisp pink radishes, herbs (basil, parsley, coriander, mint), celery, watercress and spring onions. A little salt, or even ground dried seaweed with its many other minerals, helps to mellow the bitter taste if needed, without removing its benefits.

CHALLENGES OF THE SEASON

While we may generally feel healthier and succumb to fewer bugs in summer conditions, the accompanying sun and heat can bring other challenges in the shape of dehydration, sunburn and insect bites. Luckily, there are many plant-based remedies for such things, and super-hydrating drinks to ward off dehydration.

LONGER DAYS, LESS SLEEP

June is the month with the most daylight. In the past, people were more in tune with nature, rising and going to bed with the sun. Your body clock is intertwined with day and night – as day dawns, cortisol is released to wake you up, and as night falls, your brain secretes melatonin to tell you to sleep. So, if you find you're waking up before your alarm, that's only natural. To help stay asleep for longer, use blackout blinds and an eye mask – they make a real difference. Sleep is vital to health and we need about 7 to 8 hours a night. If you're flagging, take an early afternoon nap (up to 45 minutes) to restore energy levels – any longer or later than this will affect your night-time sleep. If you have a hammock, and even someone to rock you, the rhythmic movement can induce fast and deep sleep, or even deep relaxation.

STAY HYDRATED ALL SUMMER

It's all about a balance with the sun – a little helps us make vitamin D and lifts our mood, but too much can sap energy and its heat can dehydrate us. The body is two-thirds water – water is vital for almost all body functions and is carefully controlled by our kidneys. The best ways to keep hydrated are to:

- Eat plenty of water-rich vegetables and fruit: celery, fennel, cucumber, marrows, courgettes, radishes, peppers, lettuce, watercress and other salad leaves, green leafy vegetables, melons, strawberries, raspberries; these also provide key electrolytes (dissolved minerals).

- Add mineral-rich vegetables – seaweed and nettles, for instance – to soups and salads for added electrolytes.

- Drink protein-rich fluids such as milk, nut milks, oat milk and coconut water, as the proteins and minerals hold on to water in the body.

- Include gel-containing vegetables which hold on to fluid and make it easily accessible to the body – seaweed (such as dulse and wakame) and especially okra.

- Take some of the daily recommended 6–8 glasses (about 1.2 litres) of water as herbal teas.

- Avoid sweet fizzy drinks and pure fruit juices as they are high in sugar, and prompt thirst by pulling water out of body cells.

Super-hydrating summer drinks

- Cucumber and watermelon water – blend some cucumber, including its skin, with 2 slices of watermelon, including the seeds, and add some fresh ginger or mint, to taste. As well as being delicious, there are plenty of phytonutrient benefits: the melon contains lycopene, beta-carotene and citrulline – all of which are antioxidant and anti-inflammatory – and both the melon and the cucumber are hydrating due to their mineral content.

- Fresh mint and ginger tea – put a couple of handfuls of crushed or chopped mint and a few slices or gratings of ginger with its skin into a teapot, cafetière or fridge jug and pour over a litre of boiling water. Cover straight away to capture the aromatic oils and infuse for at least 15 to 30 minutes, preferably longer, or until cool, then strain and drink. Refrigerate once cool and use within 2 days, keeping the herbs in the tea. The flavour of a long-infused fresh herb is much richer and the colour of the tea may turn dark as the pigments are released. You could make this before bed, let it cool overnight, then in the morning put it into the fridge and you will have a deliciously infused cool drink to last you through the day, or re-heat it if you prefer – this will provide nearly all your daily liquid intake.

- Kiwi and herb crush – blitz together a kiwi (with the skin if you can, though the colour will be less pretty), some ice, one or more herbs (dandelion leaves, lemon balm, parsley, lemon verbena or nettle leaf), and enough water to make a drink. Experiment with other fruits – grapes, blackberries or pineapple – and herbs, such as dill or fennel, or even some ground elder from the garden.

REMEDIES FOR TOO MUCH SUN

Marigold – *Calendula officinalis* – is an age-old remedy for burns, and sunburn is just the same. You can buy ready-made marigold-infused oil, but if you're growing marigolds for their glorious orange beauty at home, you can chop, or bruise in a pestle and mortar, 15 or so flowers, immerse them in a jar of oil (organic sunflower or olive oil), and put it on a sunny windowsill or outside for 3 weeks. Make sure the flowers are always fully covered with the oil. Then strain and press out all the oil from the flowers and use it to make the sunburn cream below.

A soothing cream for sunburn

With a hand blender, mix 50ml each of marigold oil and fresh aloe leaf gel until amalgamated. Then mix this into 100ml of base cream (be sure it's free of parabens, which can act like a hormone, and SLS – sodium lauryl sulphate, a detergent and skin irritant). Store in the fridge for extra cooling effect – it will keep for about 2 months. Apply to sunburned areas once or twice daily.

SUMMER-PROOF YOUR HAIR

The sun, the sea and the chlorine in swimming pools can all transform hair into a mass of dry, brittle locks. Avoid this by massaging some coconut oil or extra virgin olive oil into your hair after washing, wrap with a towel for 15 minutes or so, then wash out. Also, support hair growth with beneficial oils and blood circulation to the follicles by eating:

- Oily fish (up to three times a week)

- A handful of mixed nuts

- A handful of seeds

- At least 2–3 portions of green vegetables daily for their oils, as well as minerals and phytonutrients

- Red mixed berries and other dark red/purple and other colourful foods for their pigments, which promote microcirculation to the follicles

Some plastics contain strong hormone-mimics such as bisphenol-A that may disrupt hair growth and other hormonal functions, so reduce your reliance on plastics:

- Drink filtered tap water rather than water from plastic bottles

- Store food in glass or other non-plastic containers

- Use silicone covers for food, rather than clingfilm, especially if the clingfilm is in contact with the food

Such steps are good for your health, but also healthier for wildlife and for the planet, too.

REMEDIES FOR HOLIDAY WOES

According to the International Air Transport Association (IATA), over 4 billion individual journeys are made by plane – a number equivalent to half the world's population. While the average summer holiday or city break is now within financial reach of many of us, foreign trips often come with more extreme versions of our own summer environment. A few simple tips can help prevent a cold picked up on the plane, or a holiday stomach upset.

Air travel

The cabin air that dries up our nasal passages, plus being close to people sneezing or coughing, are thought to be the reasons why one in seven people catch a cold after a flight. Any potential jet lag does affect the immune system, which may help to explain the incidence of infection. In addition to staying hydrated, pack a couple of simple remedies in your cabin bag:

- Chew a sprig of an antiseptic herb – rosemary, sage or thyme – to boost your airways' defences to viruses and bacteria. The herb stimulates the local circulation and brings more white blood cells (the cells that fight off invading microorganisms) to the mouth and tonsils.

- Carry your own antiseptic throat spray – in an airport-friendly bottle – made from a strong tea of lemongrass, thyme, sage, rosemary (use a teaspoon each of lemongrass and thyme – or sage, or rosemary – to a cup of water, infuse until cool, then press out through a strainer), with an extra drop each of the essential oils peppermint and clove. Spray three or four times for one application on to the tonsils and the back of the throat, and use up to four or five times a day when travelling.

Food poisoning

You can improve your defences against infection from food or water when abroad:

- Eat more antiseptic foods – all herbs and spices for their volatile or essential oils, especially thyme and cinnamon.

- Eat the edible skin or at least the pith and seeds of fruit and vegetables in places where you trust the hygiene, to improve your gut's defences and thereby your immunity by feeding your own microbes with more fibre.

- Including plenty of omega-3-rich oils from oily fish, walnuts and other nuts and seeds may also help; lab studies show these oils can switch off the genes of *Listeria* bacteria and possibly other bacteria that are resistant to antibiotics.

Insect bites and stings

With more of our skin exposed, insects are attracted to us by the chemicals we secrete; the more active we are, the more we sweat these out. Hundreds of plants have been revealed to be sources of insect repellents – essential oils of citronella and peppermint; cinnamon leaf, black pepper, vanilla, lemon eucalyptus and shea butter. Though natural, such protection is short-lived

(3 to 20 minutes) compared to the hours offered by synthetic chemical products such as DEET. But give the natural insect repellent on page 76 a go, as well as trying out these strategies.

- Repel insects from the inside out – as well as using the known botanical repellents, eat a fresh garlic clove a day for the insect-repellent skin secretions it produces.

- Instant relief for midge or mosquito bites – chew or crush the little green chickweed, lance-like ribwort or the wider leaves of plantain to release their juice, then apply the leaf over the bite or skin and secure with Micropore tape (very useful for applying poultices like this).

- A fruity solution – fix a piece of pineapple, including its core (the part richest in bromelain, an enzyme with anti-inflammatory, antiviral and tissue healing actions), over the bite with tape to reduce swelling.

- Dress in light colours – midges, particularly, are drawn to dark colours – and if you do get bitten, resist the temptation to scratch, as this spreads the sting and sets up an itch–scratch reaction.

MAKE YOUR OWN INSECT
REPELLENT SPRAY

They may need applying more often, but home-made natural sprays are much kinder to you and the planet than synthetic repellents. Check first that this doesn't irritate your skin by spraying a little on your forearm and leaving for 24 hours. Dilute further if needed. Shake each time you use it. Once sprayed, shade yourself from sunlight if you are particularly sensitive to essential oils.

In a 100ml glass spray bottle, mix 25 drops in total of one or a mixture of essential oils –

- Lemon eucalyptus (*Eucalyptus citriodora,* citronellal-rich)
- Citronella (*Cymbopogon nardus,* citronellal- and geraniol-rich)
- Lemongrass (*Cymbopogon citratus,* citral-rich)
- Rose geranium (*Pelargonium graveolens*)

– with 100ml of water or a mix of water and aloe vera gel. This is a 1% mix (1ml of essential oil equals about 25 drops). Use a glass spray bottle, as the essential oils degrade plastic and water absorbs plastic chemicals.

An even simpler remedy to use is one based on catnip (*Nepeta cataria*). Simply make a strong tea with 50g of dried catnip to 500ml of water, leave to infuse until cool, then transfer to a spray bottle. Alternatively, mix 10 drops of catnip essential oil into 50ml of water (less than 1% dilution), shaking well before use and reapplying as needed.

SEASONAL GROWING AT HOME

With long hours of daylight and warm sun, the summer months are perfect for growing plants. Much produce can be grown from seed now and enjoyed throughout the summer. How uplifting, too, to be able to reach out to a windowbox or container, if not a garden, to snip off some fresh herbs or to uproot some radishes for instant juiciness.

A KITCHEN PHARMACY

You can grow your own kitchen pharmacy, with herbs for both cooking and medicine. Ideally, you would start sowing the seeds and plant out the seedlings in spring, but you can buy ready-grown small plants instead now. Use separate pots, if you can, to grow lavender, mint, lemon balm and aloe, as they grow fast and need more space. Thyme, basil, chives, parsley, fennel, marjoram and oregano can all be started in windowboxes, though they will eventually need to be transplanted into bigger pots. If you have to choose two, then opt for *Aloe vera* or *A. ferox* and *Lavandula angustifolia*: their beauty and multiple uses make them among the best plants to have to hand.

READY-IN-A-MONTH RADISHES

Nothing says summer more than a freshly pulled radish. There are so many varieties to choose from, and all of them produce little beauties that can be enjoyed entirely – from root to leaf. There are three main kinds of summer radish, all of them crunchy:

- The round, pinky-red radish, which is sweeter

- The French breakfast radish, more like thick, pink fingers with a white tip; these tend to be less sweet and faintly bitter

- The longer, carrot-shaped radish, which is a rarer variety

As radishes are fast-growing, they're ready to harvest 4 weeks from sowing. If you sow a batch every 2 weeks, you'll have a continuous crop that goes through the entire summer. You'll need to keep the soil moist for fleshy and juicy roots, but you can grow radishes outside in the soil or in containers, depending on what you have available.

Use the radish roots in a plate of fresh herbs and salad greens. Use the radish leaves in pesto, salads, steam-sautéd dishes, risottos and curries (which paradoxically can be cooling too, as spices like chilli and ginger make us sweat, shifting core heat to our skin). No part of the radish need be wasted.

EDIBLE CONTAINER GARDENING

So much summer produce is perfectly suited to being grown in a container – whether that container is a windowbox, large pot, growbag or repurposed crate or half-drainpipe. The key thing is regular watering.

An edible hanging basket is out of the way of slugs, though watering this is even more important.

You can still buy small or fledgling plants up to early summer and transplant these into larger containers to grow on. Some of the plants below can also be grown indoors; the Royal Horticultural Society (RHS) website has highly informative sections on growing fruit and veg at home.

- Vegetables that can still be sown from seed or transplanted are beetroot (late, for storage), tomatoes (indoors, too), radishes, potatoes, chard, lettuces, carrots, chillies (indoors), peppers, broad beans, dwarf French beans, herbs, peas, rocket, mustard greens, runner beans and spring onions.

- New to gardening? Start by sowing seeds of mustard greens (rocket, land cress, mizuna) in 15cm-deep trays and harvest within 6 to 8 weeks. Or choose a cut-and-come-again salad leaf or lettuce. Each plant should give you three or four 'cuts' before being spent.

- Easy-to-grow herbs: chives, dill, lavender, marjoram, rosemary, sage, savory, thyme, lemon balm and mint.

- Easy-to-grow container veg: tomatoes, courgettes, marrow, squash, garlic, kohlrabi, kale and artichokes.

- Easy-to-grow container fruit: blackcurrants, blackberries, raspberries, plums, damsons, greengages.

- Edible hanging basket: fill with a selection of herbs, 'Tumbling Tom' tomatoes, strawberries and lettuces.

BEAUTIFYING YOUR ENVIRONMENT

With sunny skies and rising temperatures, people usually feel better at this time of year. When windows are wide open and the outdoors and indoors mingle, there's a wealth of natural beauty to invite indoors.

EDIBLE FLOWER POWER

A host of edible flowers can beautify salads or infuse their flavour into other foods (butter, or cream for a fragrant ice cream). What's more, the flower pigments are medicinal themselves – their flavonoids support blood vessels to make them more elastic and less leaky, thereby reducing inflammation and allergy.

Edible flowers have a delicate flavour and texture. Take nasturtiums, for instance, their flowers and leaves adding a peppery hit to salads, while orange calendula petals add vibrancy and contrast to green salads.

Easy-to-grow edible flowers include:

- Marigold (*Calendula officinalis*)

- Nasturtiums (*Tropaeolum majus*), flowers and leaves

- Sweet violet (*Viola odorata*; **not African violets, which are from a different, non-edible family**)

- Heartsease and pansies (*Viola tricolor*)

- Borage (*Borago officinalis*; **flowers only; leaves are toxic**)

- Rose (*Rosax damascena* or the apothecary's rose, *R. gallica*; old roses often have the strongest perfume: *R.* 'Madame Isaac Péreire' or 'Comte de Chambord')

- Rosemary (*Rosmarinus officinalis*)

- Honeysuckle/woodbine (*Lonicera periclymenum, L. caprifolium, L. japonicum*; **flowers only; berries are toxic**)

- Sunflower (*Helianthus annuus* – best eaten as flower bud)

- Scented geranium (*Pelargonium graveolens*)

- Wild flowers – primroses (*Primula veris*), daisies (*Bellis perennis*), young red or white clover (*Trifolium pratense*), dandelions (*Taraxacum officinale*), marshmallow (*Althaea officinalis*), common mallow (*Malva sylvestris*).

Not all flowers are edible; only use those that are recommended by trusted sources (see References and Resources, page 210) and, even then, nibble them to do a taste test first.

Botanical butters

Butters infused with the scented, volatile oils of a flower or herb are a joy to look at and eat. Butter – especially from cows grazed on a variety of pasture and whose milk is high in essential fatty acids, beta-carotene and other useful phytonutrients – is healthy as well as full of flavour. The butter's oils and fatty acids support our skin and reduce inflammation. For non-dairy eaters, you can substitute coconut oil.

- Bruise or roll the petals or leaves into little tubes and slice finely.

- Mash gently into softened butter (or coconut oil).

- Chill for several hours to allow the flavours to develop.

These flavoured butters will enhance any bread, scone or cake; herb butters tend to suit savoury rather than sweet dishes.

MOOD-ENHANCING GREENERY

Research has shown that just looking at a natural view of trees, water or flowers – even if it's for as little as 5 minutes – induces relaxation, calms anxiety and reduces anger. Imagine what other benefits surrounding your-self and your home with flowers and plants could bring. The benefits of seeing and being in nature are so powerful that even pictures of landscapes, rather than the real thing, can soothe.

WELL-BEING, MOVEMENT AND SOCIAL CONNECTION

With the summer weather come countless opportun-ities to get outside more and get moving. Physical activity keeps our bodies and brains healthy and potent. In the modern world we often find ourselves sedentary. If sitting is the new smoking, then move-ment is the new longevity – walking may give us an extra seven years of life!

WALK YOURSELF HEALTHY

Being in and experiencing nature can improve our mental well-being. Parks and gardens are seen as critical to our mental health in urban areas. Also, many people are light deprived, getting as little as 30 minutes of daylight in the winter and only 90 minutes in the summer, but even on a cloudy day there's much more light outside than inside. By walking outside, one can reap the mental benefits of nature, as well as meet the challenges of light deprivation. Walking, additionally, does not have the drawbacks of traditional cardio exercise: musculoskeletal injury, joint wear and tear, elevated cortisol, muscle loss, or lowered metabolic rate. Simply put, it's the aerobic activity we were designed to do. Walking:

- Increases creativity

- Helps problem-solving

- Aids learning

Even 20 minutes of outside walking a day is useful. Get off one bus or train stop earlier. Park a little further from your office. If it takes 5 minutes to drive somewhere, consider walking and let your physical and mental health reap the rewards. Walk a dog, yours or a friend's, instead of just letting it out into the garden. Benefit even more by walking with others: spending time with family and friends and close relationships supports a healthy, longer life.

WILD SWIMMING

Plunge into cool outdoor waters to get active, feel alive and meet others – all among the wild beauty of the natural world. Outdoor swimming in lidos, the sea, lakes, rivers and pools is possible all year round, but summer is a good time to get acclimatized to the cool water.

Hydrotherapy has been used as a cure for all kinds of ailments since ancient times. Immersing yourself in cold water triggers a series of beneficial responses in the body and brain, while the act of bathing and swimming is itself therapeutic, too. The benefits of wild swimming include:

- Boosted circulation – the cold water shifts blood flow to our body's core to conserve heat and bathe our organs in fresh, oxygenated blood.

- Mental pick-me-up – it can relieve anxiety and depression.

- Stress relief – it can reduce levels of the stress hormone cortisol and raise resilience to stress and disease.

- Meditative properties – rhythmic movement keeps you in the moment.

FIND YOURSELF IN A MAZE

Mazes and labyrinths are metaphors for the human journey, places to get lost and found in. A labyrinth has only one way in and out, whereas a maze has many dead ends before the centre is found. Both offer an imaginative space in which to walk and quieten the mind, for contemplation or spiritual practice. Walking slowly along the path with the intention of clarifying a question may yield surprising revelations. You can think yourself near the centre, then find there are many turns before it is reached. People approach, then pass you. There are mazes and labyrinths all over the UK, in the gardens of great houses and on public sites. The Maze Project has a selection of incredible and unique mazes to visit in the UK.

STORYTELLING

The oral storytelling traditions passed down the generations and which bind communities are a rich and valuable part of all cultures. In fact, there is a story being told somewhere every day in the UK. Be enchanted as you are carried to other worlds and connected to others through true stories. Good stories are therapeutic, as the psychologist Bruno Bettelheim described, as they help people through difficult or traumatic experiences. Small-scale storytelling festivals now exist to bring together storytellers, musicians and poets to engage and share in stories.

CELEBRATIONS, ANNIVERSARIES AND SPECIAL DAYS

SUMMER SOLSTICE

The longest day of the year in the northern hemisphere occurs between 19 and 23 June. The solstice traditionally is a festival of growth and light while heralding declining day length. A Midsummer Day is celebrated around the world, most notably in Sweden and Norway's elaborate Festivals of Light, with dancing, maypoles and bonfires. In the UK, we could mark the solstice by making mead, a traditional drink fermented with honey. Almost any medicinal or culinary herb from wormwood to rosemary can be added to make a metheglin or flavoured mead. This is not only a truly delicious drink, albeit mildly alcoholic, but it's probiotic, too, full of beneficial microbes. The herbs in the following recipe promote health – yarrow is a digestive tonic and reduces blood pressure and allergies, while mugwort relieves gut spasm and wind and lifts mood. You can experiment and play with various edible and medicinal wild aromatic plants.

A MEAD FOR MIDSUMMER

1 jar of honey
7 jars of water
fresh yarrow flowers (*Achillea millefolium*, *Asteraceae*, daisy family)
fresh mugwort flower buds and/or flowers (*Artemisia vulgaris*, *Asteraceae*, daisy family)

Put the honey and water into a wooden or glass bowl and stir until the honey is completely dissolved. Add the yarrow and mugwort. Use your senses to gauge when enough is enough – when the smell becomes overpowering that's usually time to stop. Mix everything together well, then cover the bowl with muslin.

Stir twice a day – when you wake up and before bed. After 3 to 5 days the brew should start fermenting, depending on the room temperature.

Allow to ferment for about 14 days. Taste to see if you feel it is ready – there are no fixed rules. Remember, this is a wild herbal elixir mead, and by its very nature is unruly, feral and anarchic. It produces a deeply floral, very low alcohol (less than 4%) pre-industrial mead. Makes about 1.5 litres.

Share the mead with friends around a fire and on seasonal celebrations.

You can, once the 14 days are up, decant the mead into a demijohn with an airlock and let it sit for a month. Thereafter, the sweetness goes and it is reminiscent of a dry scrumpy (although the alcohol content is unknown at this stage!).

INTERNATIONAL FRIENDSHIP DAY

The International Day of Friendship – celebrated on 30 July – is a UNESCO initiative; many countries also mark a friendship day on the first Sunday in August. What started as a move towards fostering peace also recognizes 'the relevance and importance of friendship as a noble and valuable sentiment in the lives of human beings around the world'.

Why not use this day as a prompt to gather your

friends and their friends, to include some unknown faces, and ask everyone to bring the food in a 'pot luck' fashion – and, if possible, with local produce and especially something wild!

Such sharing of a meal, eating more slowly and stretching the time out together, adds more pleasure and health to your life. Good friends and family are now recognized as important for our health and happiness. Close relationships may also lead to a longer life. The benefits seem to be greater when we're older, because over time we let the more superficial friendships fade and we're left with the influential ones.

LAMMAS OR LUGHNASADH

This festival marks the start of harvest – on 1 August – and sits halfway between the summer solstice (the longest day) and the autumnal equinox (equal day/night length in September). It honours Lugh, the Celtic god of light, and starts off several celebrations of feasting, songs and games. Lammas stands for 'loaf mass', and traditionally people bake Lammas bread and cakes from the grains harvested then, such as wheat, barley, oats and rye. Take this opportunity to make your own bread or first find a bread-making class – get started with the sourdough starter on page 96 or the no-proving loaf recipe opposite.

A SIMPLE LOAF FOR LAMMAS

This simple, one-rising-only Grant Loaf was devised by English baker Doris Grant during World War II.

450g wholemeal, stoneground flour (spelt or rye)
1½ teaspoons sea salt
2 teaspoons dried yeast
1 teaspoon honey, molasses or Barbados sugar or malt
350–400ml warm water

In a bowl, mix the flour, salt and yeast. Dissolve the sugar or malt in the warm water.

Pour the liquid into the flour and mix well for about a minute. The dough should be elastic and soft and, when properly mixed, will come away cleanly from the side of the bowl. If it seems too wet, add a little more flour. There's no need to knead the dough, but do if you feel like it – it's quite therapeutic.

Transfer to a non-stick or oiled 1 litre loaf tin and leave to rise in a warm place for about an hour, until nearly doubled in size (it'll rise more slowly if in a cooler place).

Preheat the oven to 200°C/400°F/gas mark 6. Bake in the centre of the oven for 35 minutes. The bread is ready when the bottom of the tin sounds hollow when tapped. Turn out when slightly cooled.

VARIATIONS: To this basic recipe, you can add delicious complexities: 25–50g of medium oatmeal per 450g of flour, and some seeds such as pumpkin, sunflower, poppy and sesame; you may need a little more liquid when adding seeds. The texture will be slightly more dense but the nutritional benefits will soar; with more protein and oils, this seedy version breaks down to deliver a slow release of sugar into the blood. Other things to add include walnuts, dates, onions and olives, dried tomatoes, herbs or seaweed. Invent your own version and share the recipe with friends; give a loaf away – it's a wonderful thing to receive and great to share knowledge that can continue to be shared over and over again.

In the UK, the Real Bread Campaign lists bread-making events and classes nationwide. If you're going to give bread-making a go, why not seek out locally grown or milled grain? The UK has around 300 working mills, some in surprising places – deep in a city, such as Brixton Windmill in London – and many welcome visitors, too, during the National Mills Weekend in May.

Sourdough is an especially healthy bread to eat, but it takes much longer to make. This is part of its secret, as the long rising time contributes to the complexity of its taste and texture. As the microbes consume the simple sugars and gluten proteins, they create a texture of bread that we digest more slowly and easily, with a slower rise in blood sugar and so less stored fat. What's more, they liberate minerals, vitamins and enzymes and increase our beneficial gut microbes.

Making sourdough bread involves first making your own starter ferment, rather than using commercial yeast. Yeasts and bacteria are all around us, though usually invisible, in the air and the soil. They feed off all living things, including us, every plant surface, on grains and other foods. Adding water to flour releases sugars on which wild yeasts (such as *Saccharomyces exiguus*) and bacteria (such as *Lactobacillus acidophilus*) thrive. *Saccharomyces* produce carbon dioxide for the dough to rise, while *Lactobacilli* produce the lactic and acetic acids that make the bread taste sour – this is fermentation. Humans have used this transformative process for millennia in order to preserve foods (milk as yoghurt and cheese, grape juice as vinegar and wine, soya as tofu and miso), as the acids inhibit mould and other spoiling microbes.

- A sourdough starter – mix 30g of rye flour and 30ml of water and leave, covered with some muslin or a tea towel, at room temperature for 24 hours. The mix will expand and bubble as the yeasts and bacteria multiply and produce acids. The next day, mix in the same amount of flour and water, and repeat each day for about 5 days. By day 6, you'll have enough to use half for one loaf and keep the other half as your next starter.

AUTUMN

(September–November)

AUTUMN IS A SEASON OF TWO HALVES: abundance and conservation. As grain is harvested and fruit and nuts ripen on the heavily laden trees, nature continues abundantly. Then as the days shorten, leaves fall and plants conserve energy for the short, cold days of winter ahead. The greens of spring and summer transform into fiery reds, oranges and yellows as autumn makes itself visible. As we follow nature by harvesting, storing and preserving food, our focus moves inwards to ourselves in the cooler weather. Such shifts are a reminder that life is about continual change and that we can find a middle way, rather than all or nothing.

This cooler season is directing us to slow down and take time for reflection. To be in sync with nature, we can welcome the darker nights and consider what will make us feel good – a slightly earlier bedtime, more warming foods, and getting outside as much as possible, even for lunch or a cup of tea.

SLEEP AS THE CLOCKS GO BACK

As the clocks go back an hour in the UK in October, we gain an hour of sleep, which gives us more mental agility and protects us against stress and illnesses, such as diabetes, cancer and heart disease. It's also a chance to rest after the activity of summer and start to store our own energy for winter, both mentally for the shorter days and physically for building immunity against the infections of the cold season.

Now is the time to maximize the light you get during the day to reinforce your natural circadian rhythms, both the sleep–wake cycle and that of every body function – it will help you feel better as you're going 'with' rather than 'against' our design. As well as plenty of daylight, you need enough restorative sleep. Get into a good sleep rhythm – make the most of the extra hour of sleep when the clocks change to establish a regular bedtime and wake time, which is known to improve sleep quality.

TOP UP VITAMIN D

Since the sun's rays are much weaker in the autumn and winter (and we're not exposing our skin as much), we need to make the most of the daylight to produce some vitamin D. Like other animals, we store vitamin D in our fat and liver cells for about 2 months, so we still have some from the summer. But we can top up our levels by eating vitamin D-rich foods:

- Oily fish (such as sardines, mackerel, pilchards, herring and salmon), but also tuna and cod. Eating oily fish two or three times a week provides a certain level of vitamin D, along with high-quality protein and omega-3 fatty acids that protect our skin and our bodies against inflammatory diseases, such as heart disease, diabetes and cancer.

- Eggs, dairy milk and cheese contain some forms of vitamin D (D_3).

- Mushrooms contain vitamin D_2. Increase this by laying them on a window sill to get light for an hour or two.

In fact, many countries' health agencies now recommend a vitamin D supplement throughout autumn and winter in order to avoid deficiency and promote health.

OPPORTUNITIES OF THE SEASON

Autumn is a wonderful time to forage and get daylight at the same time. Foraging is rewarding in so many ways: it reconnects us with nature, which nourishes us mentally and physically, it is sustainable and it provides nutritious and medicinal food for free.

SEASONAL FOODS

Our bodies naturally begin to crave more substantial and warming foods as the temperature cools. It's time to make whole plant foods, especially beans, the foundation of our diet. Beans are one of our most valuable foods as medicine, as they not only provide filling, prebiotic fibre but also bring down levels of inflammatory hormones. They are harvested now for drying and storing through the winter. Squashes also come into their own, and their colourful skins are full of beneficial fibre as well as immune-stimulating pigments to sustain us through the cold weather.

WHAT'S IN SEASON?

PLANT FAMILY	EXAMPLES
Moschatel *Adoxaceae*	Elderberries
Beet *Amaranthaceae*	Beetroot, spinach, quinoa (leaves), chard Growing wild – fat hen/goosefoot, orach, sea beet, sea purslane
Onion *Amaryllidaceae*	Leeks, spring onions, shallots, onions Growing wild/herb – chives
Carrot *Apiaceae*	Fennel, carrots, celery, celeriac, parsley, dill, coriander, aniseed, caraway, fennel Growing wild – hogweed seeds (*Heracleum sphondylium*), alexanders (*Smyrnium olusatrum*), sweet cicely, ground elder, cow parsley/wild chervil, wild fennel **(caution: many wild species are toxic, so double-check identity)**

Daisy *Asteraceae*	Lettuce, chicory (endive), radicchio, Jerusalem artichokes Growing wild – dandelion, burdock root, sow-thistle, sea aster
Birch *Betulaceae*	Hazelnuts and cobnuts (*Corylus avellana*)
Cabbage *Brassicaceae*	Kohlrabi, watercress, horseradish, broccoli, cauliflower, red cabbage, Savoy cabbage, spring cabbage, kale, radishes, turnips Growing wild/herb – winter/land cress, bitter cress, hedge/garlic mustard, shepherd's purse; nasturtiums (*Tropaeolaceae*)
Hemp *Cannabaceae*	Growing wild – hops
Chickweed *Caryophyllaceae*	Growing wild – chickweed
Bindweed *Convolvulaceae*	Sweet potatoes
Cucumber *Cucurbitaceae*	Cucumbers, courgettes, marrows, pumpkins, squash

Heather *Ericaceae*	Blueberries Growing wild – bilberries, cranberries
Bean *Fabaceae*	Fresh – peas, sugar snaps, mangetouts; beans: broad, French, runner, borlotti, soy (edamame); peanuts (groundnuts) Dried – black, cannellini, black-eyed, chickpeas, borlotti, pinto, butter, adzuki, haricot, mung, kidney, soy (edamame), broad (fava); lentils: green, black, Puy, Beluga Growing wild – red clover
Oak/beech *Fagaceae*	Sweet chestnuts (*Castanea sativa*)
Walnut *Juglandaceae*	Walnuts (*Juglans regia*), hickory (*Carya* species)
Mint *Lamiaceae*	Mint, rosemary, winter savory, thyme, lavender, marjoram, oregano
Bay *Lauraceae*	Bay/bay laurel
Mulberry/fig *Moraceae*	Mulberries, figs
Sweetcorn *Poaceae*	Sweetcorn

Rhubarb *Polygonaceae*	Sorrel
Rose *Rosaceae*	Raspberries, apples, pears, damsons, plums, blackberries, greengages, peaches, quince Growing wild – raspberries, rosehips, rowan berries, hawthorn berries/haws, sloes/blackthorn berries, crab apples, bullaces, medlars
Potato *Solanaceae*	Aubergines, tomatoes, potatoes
Nettle *Urticaceae*	Growing wild – nettles
Violet *Violaceae*	Growing wild – sweet violet, heartsease, pansy
Mushrooms *Boletaceae, Agaricaceae, Physalacriaceae, Marasmiaceae, Pleurotaceae, Cantharellaceae and Meripilaceae*	Mushrooms, wild and cultivated: cep/porcini (*Boletus edulis*), chanterelle and girolle (*Cantharellus cibarius*), enoki (*Flammulina velutipes*), chestnut/portabella (mature button or common mushroom; *Agaricus bisporus*), shiitake (*Lentinula edodes*), maitake (*Grifola frondosa*), oyster (*Pleurotus ostreatus*), horn of plenty/black trumpet (*Craterellus cornucopioides*)

WHAT TO PICK AND EAT IN AUTUMN

Fruit, fungi and nuts thrive now, with timely immune-boosting components. Sprays of shiny black elderberries follow the flowers, full of antiviral properties. Rosehips are packed with vitamin C, so make the most of their goodness by storing them as syrups, jams and fruit leathers. Sloes are too astringent to eat straight from the blackthorn bush, but they add distinctive flavour – and flavonoids protecting our circulation – to gin or other spirits for drinking. Collect crab apples, the wild parent of the cultivated fruit, before or soon after they fall to the ground, and make crab apple jelly or chutney. Rowan trees (*Sorbus aucuparia*) are often found as street trees, as well as in woodlands. Like sloes, their bright red berries are too high in tannins to eat raw or stewed, so they are best made into jelly or chutney. They are higher in vitamin A than carrots, and contain significant vitamin C, pectin and organic acids. The berries lower blood pressure and support circulation and bowel function.

Elderberries

The elder tree is a generous one – first we can harvest its flowers for cordial and all manner of scented treats, and next its berries offer antiviral action.

- A sweet way of preserving these berries is to make potted elderberries. Strip elderberries from their stalks with a fork. Weigh the berries and allow 40g of cane sugar or honey per 450g of berries. Mix the sugar (or honey) with the berries and fill glass preserving jars. Cover the jars with their lids, and put into a preheated oven (160°C/325°F/gas mark 3) for about an hour, until the juice has run fully out of the berries. Top up one jar from another if needed and cover immediately. The berries will keep for months or even years. Add a teaspoon to breakfast smoothies and desserts to prevent or treat coughs and colds.

- Another easy way to have a year-round supply of these berries is to dry them after harvesting in autumn. Use a fork to remove the berries from the stalks, then spread them out on trays, put them into a very low oven (50–90°C/120–195°F/gas mark ½), and leave them for several hours until they shrivel completely. Store them in a tin, china or glass jar and label with the date and place of harvest.

- The dried berries make an excellent warming antiviral tea that you can drink throughout the day. Simmer 3 teaspoons of berries per 3 cups of water for 10 to 15 minutes, covered, with some slices of fresh ginger and 2 cloves; add some honey, to taste.

Mushrooms

Identifying and learning how to find and use mushrooms is a world in itself. They are a source of protein (around 20% when dried), vitamin D and, more importantly, their beta-glucan polysaccharides (a form of carbohydrate) stimulate key immune cells, which respond to viruses and bacteria. Even the humble button mushroom (*Agaricus*) contains beta-glucans. So mushrooms are a key food during the autumn, both to protect against and to treat us for colds and other infections. You can buy so many types of mushroom now: chanterelle, enoki, chestnut (portabella), shiitake, mitake, girolle and oyster, among others. There are also **non-edible mushrooms such as the beautiful turkey tail (*Coriolus versicolor*)** that grows in woodlands and reishi (*Ganoderma lucidum*); both are commonly used in plant medicine. Find a local mushroom expert to explore these extraordinary organisms. If you want to learn about identifying mushrooms, look for a qualified guide with long experience and public liability insurance.

CHESTNUT AND MUSHROOM SOUP

Sauté a chopped onion and a chopped stick of celery in olive oil and/or butter until soft. Add 150g of sliced mixed mushrooms (including shiitake or other exotics) and 1 litre of stock, e.g. chicken or vegetable (or 1 litre of water and a handful of seaweed), and simmer for 15 minutes.

Add 180g of boiled or roasted chestnuts (vacuum-packed are fine) and simmer for a further 10 minutes, until all the vegetables are soft. Season with salt and pepper, then purée. Add a little crème fraîche for a final flourish to this velvety soup. Serves 4.

VARIATIONS: You could substitute root vegetables, such as celeriac or Jerusalem artichokes, for the chestnuts, or simply add these vegetables with a bit more stock to the mushroom/chestnut mix.

Pulses

The bean or legume family (*Fabaceae*) is large, containing around 1,800 species, some with glorious colours and patterns, but we only see a small selection. Pulses are dried beans and lentils; they contain about one-third protein and are high in protective phytoestrogens (plant oestrogens) that keep our own powerful hormones oestrogen and testosterone in check. Their protein and fibre keep us full and they release their sugars slowly, thereby helping to prevent fatigue, weight gain and mood swings. Their fibre improves our bowel habit and feeds the beneficial gut microbes that make energy, vitamins, nerve chemicals and more.

If you can, buy British-grown pulses, which are becoming more available. These would be excellent to sprout, being relatively fresh.

SPROUTING YOUR OWN

Sprouting is the germination of the seed and is a way of making raw pulses edible and digestible. It can be done with most seeds, though the potato family (*Solanaceae*) should not be eaten as seeds. Follow these steps to grow your own sprouts.

- Take a large glass jar and make small holes in the metal lid or stretch some muslin over the top and secure with an elastic band – this is your sprouter.
- Fill about a third of the jar with beans or lentils, rinse them with water, then fill the jar with lukewarm water and soak overnight.
- Next, rinse and thoroughly drain them by turning the jar upside down. Lay it on its side in a dark place. Keep it at room temperature, rinsing and draining the contents two to four times a day.
- The seeds should fully sprout in 3 to 5 days. Drain them well and put them into the fridge to use within 2 or 3 days.

THE ULTIMATE PULSE-GRAIN DISH

This is the daily go-to dish, not only for autumn, but beneficial for us when eaten all year round. It provides the base that balances our energy and hormones through the day, while providing some medicinal polyphenols. Cook once, on a

Monday morning, and you have a dish that keeps for 4 days, providing 4 instant lunches. The beauty of this base dish is its flexibility – by varying the bean and grain used you can create a wide variety of different meals from one recipe.

Mixing a pulse and a grain gives you complete protein – with all eight essential amino acids. Beans and rice is a dish famous around the world in many guises – Caribbean rice and peas, Asian rice and dhal, the Middle East's mujadarra.

Eat a few tablespoons every day for lunch, along with a few simple additions of whatever you have handy – oily fish (you can use tinned), hard-boiled egg, feta, watercress, a few baby plum tomatoes or some roasted vegetables. The key to bringing it all together is a tart dressing of extra virgin olive oil or other cold-pressed oil, some balsamic or cider vinegar or lemon/lime juice, and salt and pepper. You can also add vegetables while it cooks.

½ cup of black, Puy, Beluga or green lentils (better than orange lentils, which have been skinned)

½ cup of wholegrain rice: wild, black/red, short-grain brown or brown basmati; barley,

spelt or freekeh; buckwheat; millet, amaranth or quinoa
2–3 cloves of garlic, skin on
wakame, sea salad or other mineral-rich seaweed/sea vegetables, chopped
2 bay leaves
a few peppercorns
3–4 slices of fresh ginger, skin on
optional: a handful of chopped mushrooms
a little salt, pepper, extra virgin olive oil, lemon juice
a large handful of chopped parsley or coriander

Put roughly equal amounts of lentils and wholegrain rice or other grain into a pan with the garlic, seaweed, bay leaves, peppercorns, ginger and mushrooms (if using). Add about double the volume of water, about 2 cups, then cover and simmer very gently (a heat diffuser is helpful) for about 35 minutes. Check halfway through that the water has not boiled away and add a little more if necessary – it should be completely absorbed once the lentils/grains are cooked. Then season with salt, pepper, oil and lemon, and add the chopped parsley or coriander for extra minerals. Keeps in the fridge for 4 days. Serves 4–5.

> **TIP:** Start with small amounts of beans, lentils and wholegrains to allow the bacteria in your gut (your microflora) a chance to multiply to digest their fibre, and any increase in wind will soon go back to normal.

Horseradish

Part of the mustard/cabbage family, horseradish (*Armoracia rusticana*) is one of the most revered traditional remedies for colds and chest infections. You can harvest a horseradish root at any time of the year, but it's tastiest when its edible leaves die down in the autumn and the sugars move into the root for storage. If you're considering planting horseradish at home, know that it is tenacious and spreads – not even a very hard frost will kill it. A supermarket-bought root may grow if you plant it. If you don't have space to plant one, ask a local allotment holder if you might dig up some root, as this helps to control its growth.

- Peel, then grate the outer, softer flesh and compost any woody part. Mix into yoghurt

or cider vinegar and take a teaspoon a day
when you have a chest infection.

- Or mix into crème fraîche and serve with
  beetroot, salmon or mackerel.

Do eat it raw, as cooking destroys the therapeutic
effect of its volatile oils. Grating horseradish will help
clear congested sinuses.

Nuts

Hazelnuts and cobnuts (a cultivated variety of hazel-
nut, *Corylus avellana*) are rich in protein, oils and
minerals. They are seen in Celtic mythology as giving
wisdom and protection. Displayed in their clusters on
a plate with some apples, they lend a beautiful, seasonal touch to a meal. Sweet chestnuts, too, are an
autumnal joy – high in fibre-rich carbohydrate, vitamin C, B vitamins and slow-release energy, either
roasted or combined with mushrooms or meat dishes.

HAZELNUT BEAN BEETROOT BROWNIES

Here is a rare thing – a sumptuous and healthy
brownie recipe! It avoids the typical brownie
sugar rush, as it contains protein, fibre and
mineral-rich dates instead of refined sugar. Its

beans, beetroot, chocolate and nuts balance the butter or oil and provide a range of valuable phytonutrients to support mood, immunity and circulation and balance our hormones.

200g dark (70% cocoa solids) chocolate
200g unsalted organic butter or olive oil
150g pitted dates
3 eggs
150g cooked black beans
200g raw beetroot
150g ground hazelnuts (best toasted)
1 teaspoon baking powder
optional: crushed hazelnuts

Preheat the oven to 180°C/350°F/gas mark 4.

Melt the chocolate and butter or oil in a bain-marie or a large bowl over a pan of water. Blend the dates and eggs until smooth. Process or mash the beans until almost smooth but still with some texture. Grate the beetroot into a large bowl, then add the ground nuts, baking powder and all the other ingredients apart from the crushed hazelnuts. Mix everything together well by hand.

Line a 25cm round cake tin or similar sized brownie tin with baking paper and pour in the

mixture. Sprinkle over some crushed hazelnuts
if you like. Bake in the preheated oven for about
25 minutes. Don't overbake – the centre should
still be a little moist. Leave to cool in the tin,
then tip out and cut up as you like.

Try out different versions of this recipe using
other root vegetables, nuts, beans or even seeds
and oils such as coconut oil.

Rosehips

Used worldwide for medicine and food since ancient
times, rosehips are the red jewels of autumn. During
World War II rationing, the UK's Ministry of Food
encouraged children to pick rosehips for home-made
syrups to ensure they had enough vitamin C when
other sources were scarce. It is uplifting to pluck these
gorgeous red hips from the plant and transform them

into tea, soup, marmalade, dried fruit pulp, syrup, purée and fruit juices. In Scandinavia and Turkey, people use rosehips in fruit soups.

You can use the hips from any pesticide-free roses – including those in your garden. Hips from wild roses provide not only one of the richest sources of vitamin C but are full of other medicinal benefits for the heart, circulation, skin and digestion. Our tough native species include dog rose (*Rosa canina*) with its pale pink flowers, field rose (*R. arvensis*) with white flowers, and the pink, apple-scented sweet briar or eglantine (*R. rubiginosa*). All produce clusters of bright red, oval hips. The Japanese beach rose (*R. rugosa*) bristles with thorns and bears large, orange-red, flatter hips from late summer, though these tend to be less flavoursome.

If you don't fancy a foraging trip, you can buy dried rosehip powder from herb suppliers – though its vitamin C levels will be lower, the rose's many other phytochemicals remain, so it's a powerful medicine. Check it smells fresh, as powders have a short shelf-life. Just add a teaspoon daily to fruit smoothies and any other dishes.

ROSEHIP PURÉE

This is the simplest way to get the most flavour and colour from rosehips. First, gather the hips when they are a rich red, ideally after a frost when they soften naturally. Nibble to taste them for sweetness and vitamin C acidity. If they are red but a little hard, soften them by putting them into the freezer and then defrosting them.

Remove any green stalks and blend them with enough water to make a purée. Strain the purée through a double layer of muslin to remove the hairs and seeds. Keep the purée for up to 3 days in the fridge or freeze in ice-cube trays. Add 1–2 tablespoons, or a cube (defrosted if needed), daily to yoghurt, breakfast muesli or a smoothie.

TIP: Pluck hips from plants growing on higher ground, as they are likely to contain more vitamin C.

CHALLENGES OF THE SEASON

The dry, warm air of summer months is cooling but the dry, now-cold air can dry out the mucus membranes lining our nose, throat and lungs. Along with specialized hairs, this mucus is designed to trap and expel irritants such as dust particles, viruses and bacteria. These autumnal atmospheric conditions allow coughs, colds, flu and other viral infections to take hold more easily; evidence shows that flu epidemics usually follow a drop in air humidity.

When we cough or sneeze, we expel a mist of particles from our nose and mouths. When the air is moist, these particles remain relatively large and drop to the floor. But in dry air they break up into such minute parts that they can stay aloft in a room for hours or days, so we may continue to breathe in a cocktail of dead cells, mucus and germs.

BOOST PRE-WINTER IMMUNITY

We can build more resistance to airborne infections by shoring up our immune defences with basic self-care and the following steps:

- Take a vitamin D3 supplement (400–1000IU) every day; the dose depends on how much daylight you get.

- Stabilize blood sugar and insulin levels by eating slow-release wholefoods at breakfast and lunch; include a portion of beans or lentils as a daily dose of anti-inflammatory power.

- Increase your daily intake of colourful, antimicrobial and immune-stimulating foods (including mushrooms, black/red berries and purple potatoes for their protective polyphenol pigment, ginger, turmeric, cinnamon, cardamom, thyme and sage, along with healthy fats such as oily fish, nuts and seeds and cold-pressed oils).

- Enjoy hot, pungent or essential-oil-rich foods and herbs – chilli, ginger, mustard, horseradish, garlic, basil, thyme, rosemary and sage – for their antiseptic properties and ability to thin mucus so that it can more easily be coughed away.

- Check where to find local elder trees and wild roses – these bear the most antiviral and immune-stimulating berries for coughs, colds and flu. When they are black and ripe, pick a big bowlful of elderberries to turn into a spicy, medicinal syrup or to bottle. The same goes for rosehips. If you don't have time to make remedies with them straight away, simply freeze them.

- Replace water loss from catarrh with warming teas or water.

- Humidify and purify the air in your home by placing saucers of water with a few drops of air-antiseptic essential oils, such as lemongrass, eucalyptus, rosemary and thyme, near radiators, or use electric essential oil vaporizers.

- Carry a spray bottle to use as an air antiseptic – mix a 1% concentration with 7 drops of essential oil of tea tree (*Melaleuca alternifolia*) and 7 drops of eucalyptus (*Eucalyptus radiata* or *E. globulus*) in 50ml of water; shake very well before use to disperse the oils, as they don't mix well in water, and give two or three spritzes around you when in company (but not into eyes) – others should benefit too!

SIMPLE REMEDIES FOR SORE THROATS, COUGHS, COLDS AND FLU

Many plants have antiviral and antibacterial actions, through their volatile oils and other phytochemicals, that we can use both to prevent and to treat infections and also to stimulate our own immune defences.

Antiviral tea

Start drinking this medicine now, before the cough and cold season is well under way. It's best made the day before (for the full medicinal components to infuse the tea); ideally, make enough tea for a whole day and take to work in a flask or drink cool.

- Simmer 3 teaspoons of dried elderberries or rosehips, 3 crushed cloves and half a crushed stick of cinnamon in a pan with 500ml of water for about 20 minutes.

- Leave the berries or hips and spices in the tea to cool overnight, then strain and drink, adding a little honey to taste. Reheat the tea if you prefer.

- Other antiviral or antimicrobial herbs to enjoy as teas are lemongrass, thyme, rosemary and sage.

Quick throat gargles

Gargling is a way of applying antimicrobials directly to the infected throat and tonsils. The essential oils stimulate these tissues to boost blood circulation and production of immune cells. If there is no pus produced by the tonsils, you can swallow the gargle; otherwise, spit it out. Use at least 3 or 4 times daily in an acute infection and continue until the infection has gone.

- Mix a crushed garlic clove and 1–2 teaspoons of lemon juice with ½ a glass of water and a dash of honey, if needed.

- Make a strong tea with 2 teaspoons each of crushed fresh thyme and sage in a mug – pour over boiling water and cover immediately to trap the antimicrobial essential oils. When cool, strain and gargle.

Vapour rub

Applying this pungent balm with diluted essential oils to the skin promotes circulation, decongestion and antimicrobial action. The oils reach the blood and lymph circulation below the skin. It keeps in the fridge for several months.

- Mix a total of 20 drops of one type of essential oil or 5 drops of each of the following essential oils: rosemary, peppermint, pine and/or eucalyptus (*Eucalyptus radiata* is more antiviral than *E. globulus*) with 30–40ml of a base oil (such as light olive or sunflower oil) or into 30–40g of a base cream (such as E45 cream).

- Rub some of the oil or cream into the neck, chest, armpits and behind the ears, but not near the eyes. Apply twice daily (once in the morning and once at night) and, where possible, cover with clingfilm to increase absorption of the essential oils through the skin and into the lymph nodes and circulation beneath.

Medicinal honey

This powerful medicine in a jar lasts up to a year, so you can dip into it whenever you feel a sore throat or cold coming on.

- Fill a jar with unpeeled garlic cloves or sliced onion along with at least 2–3 tablespoons of crushed fresh thyme (for its strong antimicrobial action) or a good tablespoon of powdered turmeric.

- Top up with honey to the brim of the jar, to prevent spoilage from reaction with the air. Seal, label and date; it needs to sit for at least 2 weeks before it's ready to use.

- Take by the teaspoon (up to 5 teaspoons a day) to soothe colds, flu, sore throats and coughs.

Liquorice cough syrup

- Simmer 2 tablespoons of chopped liquorice root in 500ml of water, covered, for 20 minutes. Remove from the heat, add 2 tablespoons each of chopped thyme and sage, cover again and infuse for another 20 minutes. Strain the liquid, add 200g of honey (or raw cane sugar) to it, and simmer, stirring, until syrupy, then cool. Pour the mixture into labelled bottles and use within 6 months.

- Take 1–2 teaspoons neat, as required.

SEASONAL GROWING AT HOME

In autumn we are still reaping the benefits of summer-sown salads and late harvests of tomatoes and runner beans, while starting to sow seeds for next spring. Several vegetables are sown in the early autumn to allow harvesting in early spring, including broad beans, garlic, lettuces, onions (and spring onions and shallots). Early peas, winter spinach and spring cabbage can be sown outdoors throughout September to harvest in March and April. Sowing seeds in cool weather when nature is falling asleep feels like an act of hope, with the promise of young seedlings to look forward to in the spring.

INDOOR MINT

As one of the most vigorous herbs, mint is easy to keep growing indoors by taking root cuttings. Fork gently around the plant to lift some of its copious, whitish roots. Cut off a few 5cm lengths, lay the roots on the surface of a pot of compost and cover with a little more compost. Water, then place on a window-sill indoors and new shoots will soon appear.

HEALING FLOWERS TO SOW NOW

The all-important healing pot marigold (*Calendula officinalis*), with its brilliant orange face, is a hardy annual, meaning its seed will withstand frosts and snow to flower the following year. Sowing these seeds now will mean you can harvest the flowers in the spring to make a cream or ointment for all kinds of skin wounds, fungal infections, eczema and sunburn (see page 70). Another cheery, orange annual and easy to grow, though less hardy, is the Californian poppy (*Eschscholzia*

californica), which has calming properties and is used to promote sleep. The red or corn poppy, *Papaver rhoeas*, will also grow well in a pot – its tissue-paper-like flowers are also used for insomnia. When they flower in late spring, make a tea of their flowers (put 2 generous teaspoons of fresh flowers (or 1 teaspoon of dried flowers) into a cup or mug, cover with boiling water and infuse for at least 15 minutes. Strain and take to bed as a sleep-inducing tea. The seeds of both plants are delicious and edible and are often used in cooking. Unlike the addictive alkaloids in the opium poppy, *Papaver somniferum*, the alkaloids in these poppies are much gentler and safe to use even for children.

PLANT BULBS NOW FOR WINTER SCENT

Coming home to the intoxicating scent of narcissi or hyacinths is one of the joys of winter to prepare now. Planting bulbs in a container if you don't buy them ready planted is so easy. Favourites are the beloved daffodil variety known as 'paper white' (*Narcissus papyraceus*) and any hyacinth (*Hyacinthus orientalis*), with pink, white or blue flowers, for their potent fragrance.

Although their natural flowering time is spring, if you plant them indoors in autumn they will flower around Christmas. Each bulb costs little, but gives much back – just one flower in a glass by your bed and in each room will infuse your entire home with a delicious perfume. What's more, bulbs are perennial, so after flowering, either plant them straight into garden soil, or keep them to replant indoors next year. Clean off any soil and trim the roots and any flaking outer layers of the bulb. Lay them out to dry for 24 hours or so, clean off any soil, then put the bulbs into paper bags and store in a dry, cool place.

GROW YOUR OWN GRAPES

If you have a balcony or garden that faces south or gets plenty of sunshine, then plant a vine (between October and March) – this will not only give you a host of pleasures and health benefits but will also allow you to enjoy the grape in its entirety – its fruit, seeds and leaves.

Grape skins and their seeds contain some of the most powerful natural medicines. Why buy expensive grape-seed extract when we should have the choice of seeded grapes in supermarkets? It's a human right to have food with its own medicine intact as nature intended. This is how we thrive.

Red and black grape skins, for instance, contain resveratrol – an anthocyanin (polyphenol, a phyto-nutrient) pigment, also present in red wine, designed to protect the grapes against infection, stress and UV light. This appears to have multiple benefits for us – it's antioxidant, anti-inflammatory, a brain support, anti-diabetic, supports gut microbe function and may help prevent or treat dementia. Resveratrol is also found in many other foods, including Japanese knot-weed, peanuts, pistachios, blueberries, cranberries, and cocoa and dark chocolate.

- Plant grapes in a container, greenhouse or garden.

- Snip off unripe green grapes and press or juice them to collect their sour verjuice, which is a delicious alternative to lemon juice and vinegar (high in protective polyphenols), to flavour salads and sauces.

- Use the leaves as a lining for tarts and pies. Blanch them briefly in boiling water, then line a baking dish with them so they generously overlap the sides. Fill them with a mix of wilted spinach or wild greens, thick strained yoghurt (*labneh* in Middle Eastern cooking) or goat's/sheep's or cottage cheese, pine nuts and beaten egg, cover with the vine leaves and bake.

BEAUTIFYING YOUR ENVIRONMENT

This is a good time to encourage wildlife on to a balcony or into your garden when they most need protection from the cold.

BRING WILDLIFE INTO YOUR GARDEN

Encouraging a plethora of creatures – be they beetles, birds, bees, bats or hedgehogs – into your garden or outside space brings an instant connection with nature.

- The best – and easiest – way to feed wildlife in a garden is to mimic what nature provides and let part of your green space go a little wild – make a small pile of stones or a rock garden for red mason bees; a pile of wood for beetles and woodlice; a pond, as overleaf; and cut off the lower part of the bottom of a fence for hedgehogs, as they like to roam.

- Simply putting some water in a bowl and some snacks out in the cold weather will support birdlife.

By doing a little to help these species and their links, we maintain this chain of life and keep the whole environment alive. We benefit, too, from the sight of birds and the joy of their song, and the sheer variety and extraordinary design of insects and other animals.

MAKE AN EDIBLE POND

If you tried growing watercress in a bowl (see page 27), go one step further and create a pond. In the UK, pond numbers have declined by about 80% over the last century, due to pollution and changes in farming practices. But ponds can support an astonishing two-thirds of all freshwater wildlife, and together they create a corridor to help pondlife move from one wet area to another, helping species thrive. Even a small pond of 1 square metre contributes significantly to the diversity of life, as you'll be supporting insect life – think dragonflies, aquatic beetles, mayflies and caddisflies – as well as endangered amphibians, such as the great crested newt.

The flower buds and petals of waterlilies (*Nymphaea alba*) are edible, as are the gingery-tasting rhizomes (root type) of sweet flag (*Acorus calamus*). Use these rhizomes as a substitute for ginger, cinnamon or nutmeg in cooking; in the past, they were candied and used as a sweetmeat. The inner portion of the young stems can be eaten raw and the young leaves can be eaten cooked. The mature leaves of sweet flag are insect-repellent, and the lower stem and rhizome can be dried and used to scent clothes and cupboards. Sweet flag is a well-known herb used for reducing stomach acidity and wind. Like all wild plants, these water plants contain more phytonutrients and antioxidants than cultivated species, so are worth adding in small amounts to your diet.

COLLECT AND DISPLAY LEAVES AND FRUITS

As autumn progresses, the magnificent colours of the leaves change almost daily from green, through oranges, yellows and reds, to their ultimate shades of brown. Fruits, too, are a joy to behold. Take quinces, for instance; pick them when they are golden, store them for a while on a plate, not touching, and enjoy their unique scent and irregular shape. They are edible when cooked, with a taste quite distinct from apples or pears. Cook them very slowly in the oven or on the hob with sugar – you can also add them to meat, game or vegetable stews.

Sculptural plants such as wild teasels or garden alliums, and sprays of gorgeous orange-red tree leaves – maple, beech and oak – are wonderful in vases, simply arranged. If you're lucky enough to have a grapevine in your garden, pick its green, citrus-tasting leaves and use them instead of pastry for a savoury tart (see page 130), and later, when they turn red, use them to line a bowl or plate for cheese or fruit.

WELL-BEING, MOVEMENT AND SOCIAL CONNECTION

As daylight reduces, we need to make the most of it in order to reinforce our body clock and keep our spirits up and our bodies in good health. When you are walking and exploring, notice which birds are on the wing or seeking food, how they move, and hear their sweet songs.

Discovering hidden places is another part of the pleasure of getting outdoors. To learn more about the plants, stories and events that make your local area distinctive, Common Ground is an imaginative place to start. Its Local Distinctiveness Rules page lists many ideas, and its website features various projects that reconnect people to a place and its particular history.

When you're out, wrap up warm, as our resilience to infection decreases if we get cold for long periods. However, short bursts of cold or shivering can strengthen our resistance to cold. If you started the invigorating pastime of wild swimming in summer, continuing to swim through autumn and winter in cool waters, once you've acclimatized to them, brings many health benefits.

STEP OUT INTO THE DAYLIGHT

We evolved over millions of years to eat in daylight and sleep with the dark, but electric light in the evening has shifted our eating and sleeping habits later, which is stressful for the body.

So it's very important to expose yourself to natural

morning and afternoon light. This helps reset your circadian rhythm and is easy to do. Consider walking to work or even just to the station instead of getting a bus, hold meetings with friends or colleagues outdoors, and have lunch or your morning coffee outside, well wrapped up against the cold – all these are ways of absorbing valuable UV light. Such exposure has been shown to ease depression, boost mood and overall health and encourage falling asleep faster and better-quality sleep.

One of the rewards of going for a walk near your work or home is discovering hidden secrets of peace or beauty – a little churchyard, tiny park or other corners unseen before – a refuge from the stresses of life or a quiet place to share lunch with a friend. There's probably much more to your neighbourhood than you know. A local history or wildlife guide may help to further reveal the unseen history or life of cities, villages and the land.

BONFIRES AND OUTDOOR PARTIES

Among the joys of autumn are the smells of bonfires, roasting chestnuts and hot drinks with friends outdoors at night. Relish such celebrations and minimize your impact on the environment and local wildlife by following these tips:

- Go to a public bonfire and firework display, rather than having your own; if you'd prefer your own bonfire at home, keep it small to reduce the carbon created from burning wood.

- Before you light the fire, check that there are no hedgehogs or other small creatures sheltering underneath.

- Use kindling, rather than firelighters or petrol, and untreated, dry wood to create less smoke and less irritation for the lungs.

- Best to avoid releasing balloons and lanterns into the sky, as they cause not only litter but also death or injury to wild birds and animals; in fact, many councils in the UK have banned them.

OUT IN THE NATURAL WORLD

Taking part in one of the local wildlife trusts' volunteer clean-up days – be it for woodland, beach or river – creates opportunities for further discoveries as well as connection with your local environment. And wherever you go, think about the drive to pick up plastic – take three pieces of rubbish when you leave the beach, park, waterway or . . . anywhere!

HOW TO CREATE EMOTIONAL RESILIENCE

Plants and trees offer models to remind us of the wisdom of the human body. The flexibility of an oak or a reed, with the capacity to stand its ground while bending to huge winds, can encourage us. Resilience is something we can build, with practice. We can build emotional resilience through gratitude, altruism and a sense of purpose. Reaching out to help other people is a way to move outside ourselves and enhance our own strength. It's about taking responsibility and creating a life that is meaningful and with purpose. Undertaking challenges – climbing a mountain, taking an adventure holiday, wild swimming or sharing a skill – can involve positive stress, by stretching ourselves through achievement. These experiences build resilience to enable us to handle stress better. Stress will never go from our lives, so find ways to respond to or rest from it – take a walk, talk to a friend, spend 5 minutes with your eyes closed or meditating.

WRITING MORNING PAGES

American author Julia Cameron describes how artists can develop inner awareness to help them tap into their unique inspiration by writing what she calls 'morning pages'. It is simply writing down in a note-book your thoughts, dreams or whatever is on your mind as soon as you wake, with no literary skill needed, as this is for your eyes only. Just keep a note-book and pen by the bed. The process of getting thoughts straight out of your head and on to paper is a surprisingly effective way to learn more about yourself and help find direction. It is also a great way of offloading thoughts that may keep you awake in the middle of the night.

CELEBRATIONS, ANNIVERSARIES AND SPECIAL DAYS

The autumn equinox (21–23 September) marks another date in the calendar when the length of the day and night is roughly equal. The idea of a festival in honour of a bountiful harvest is found throughout the world in many different cultures – from Korea's three-day festival of Chuseok and America's Thanksgiving, to Finland's Strömming, which celebrates a successful end to the Baltic herring-fishing season.

APPLE DAY

A symbol of harvest and fertility, the apple is treasured in many sacred traditions. It stores well, its crispness and sweetness cheering us through the cold months. Responding to the gradual loss of apple orchards in the UK, Apple Day on 21 October was established in 1990 by Common Ground to celebrate local communities and their many types of apples. Granny Smith and Braeburn are among the most medicinal varieties and are commonly available, while more unusual and traditional varieties, including Ribston Pippin, Dumelow's Seedling and Adam's Pearmain, are sold at farmers' markets.

Apple Day has since led to the rediscovery of many varieties thought to be extinct. It has also inspired Damson, Plum and Pear Days. Orchards used to be places for social gatherings as well as sheltering a wealth of animals, insects and other biodiversity. At Apple Day events, an expert is usually on hand to

identify species and advise on apple growing, along with tasting, juicing and baking. It's a chance to share recipes, learn about wassailing (drinking to the health of your fruit trees as well as your companions) and find out how to make your own cider. Often there are activities, from simple apple printing to mummers' apple plays, new songs and poetry evenings. All this brings us closer to the places we live in and helps us understand the provenance and traceability of food. These events also show that we are not alone in valuing the links between food and the land.

APPLE PIPS

The seeds of the apple's core have been unfairly demonized. While the seeds contain amygdalin, a compound that breaks down to cyanide on chewing, a 59kg adult would have to chew at the very least 176 and possibly nearer 5,000 seeds for a fatal dose. So eat your apple core and its pips without fear, as they are also a rich source of beneficial polyphenols. Amygdalin is a part of many food plants, including beans and linseeds. It has been studied as an anti-cancer drug, though conclusive evidence of its effectiveness has not yet been drawn.

FRUIT LEATHERS – THE PERFECT POCKET SNACK

These dehydrated treats are delicious, and a concentrated fruit snack and medicine in one. They're easy to make – just start the night before and they'll be ready the next morning.

Preheat the oven to 70°C/160°F/lowest gas mark) and line a baking tray, lightly oiling the baking paper.

Roughly chop a mix of apples, blackberries, rosehips and other berries, including all the skin, pips and cores, and transfer to a saucepan. Give it an immune-support boost with some spices (ginger, cloves, star anise, allspice, cinnamon or nutmeg), and pour over a cup of water for every 8 cups of fruit. Cover and simmer gently for about 15 to 20 minutes, stirring occasionally, until the mixture is soft and pulpy. Add a little sugar or honey, to taste, if needed.

Depending on the mix you've used, either sieve, or whiz in the processor then sieve until you've got a smooth purée. Spread the purée evenly on the prepared tray and transfer to the preheated oven for 12 hours, or until all the moisture has evaporated.

Cut into strips with the paper underneath and roll tightly to store in a tin for about a month – best in the fridge, or for up to a year in the freezer.

HALLOWE'EN

Hallowe'en (31 October) is All Hallow's Eve, the day before All Saints' Day, which is then followed by All Souls' Day. Traditionally, bonfires were lit to mark the end of summer and symbolize purification, with a night of fasting, rather than the mischief and magic of today. Now, apple bobbing, cider, and pumpkin carving go alongside trick-or-treating.

All Saints' Day celebrates Christian saints and martyrs, and All Souls' Day and the Celtic equivalent, Samhain or Feast of the Dead, is an opportunity to remember all our ancestors. 'Souling' would take place the night before All Soul's Day, as 'soulers' went door to door begging for spiced ale and soul-cakes made with fruit, butter and spices in return for prayers and songs. The Spanish celebrate All Saints' Day with their matchless *pannellets*, little almond cakes studded with pine nuts, and street vendors roast chestnuts and sweet potatoes.

Pumpkins and squashes – now synonymous with Hallowe'en in many countries – also bring with them vivid colour and top nutrition: they are rich in beta-carotene and vitamins B6, C and E, as well as magnesium, potassium and manganese. Their sweet-tasting flesh takes on seasonings well and they will store for weeks or months. Plus, their skin is edible – just soak them first for 5 minutes with a teaspoon of bicarbonate of soda dissolved in the water to get rid of any chemical residues.

But how to use up the thousands of tonnes of pumpkin thrown away by most of us after making lanterns at Halloween?

- Eat the seeds, roasted with the unskinned flesh (the skin and seeds are the most valuable parts!) or separately with a little salt and olive oil – they're free here but expensive to buy and are full of zinc, other minerals and vitamins and oils.

- Make soups, risottos, purées with beans or chickpeas for dips, and add to salads either grated raw or roasted.

- Add water to the seeds and stringy flesh and simmer for 10 minutes, then strain and mull with cider, cloves and cinnamon. Or use it as a stock for soups.

- Make pumpkin purée, also known as pumpkin butter; it's mashed pumpkin, including the skin, cooked with apple juice, maple syrup, cinnamon and allspice, then puréed and stored in jars.

- Freeze the flesh if you can't use it immediately.

- Save some seeds to grow more pumpkins.

- Wash or cut out any soot from the left-over pumpkin lantern and roast or add to soup.

A QUICK PUMPKIN AND SWEET POTATO TORTILLA

Grate sweet potatoes and/or pumpkin with their skins and add to briefly steam-sautéd leeks and kale. Add beaten eggs with salt and pepper and cook in a frying pan on both sides, as you would a tortilla or a griddle cake. If you like, add mashed chestnuts and cheese to the vegetables for an extra savoury hit.

DIWALI FESTIVAL OF LIGHT

Falling in November (the date varies), Diwali is one of the most significant festivals in Indian culture. Signifying a row of lamps, Diwali originated as a harvest festival and is now celebrated for many reasons, including the triumph of light over darkness, and good over evil. Every autumn millions of Hindus, Sikhs and Jains across the world attend firework displays, decorate and light their houses with colourful lights and candles, feast and share gifts. Diwali's health-giving rituals include waking early to light lamps and aid the body's overnight cleansing, the breathing practice of *pranayam*, applying fragrant massage oils and powders, and taking warm baths with rose petals and other perfumes.

Sweetmeats are central to the festival when the season is becoming cold, dry and windy. Applying the nutritional principles of Ayurveda, India's traditional medicine, Diwali dishes and sweets should help the body manage the cold season with a balance of warm and cool ingredients. The following recipe is a version of *ladoo* or *laddu* – sweet sesame balls, with warming seeds and nuts and nourishing butter or oil. The sesame seeds are high in minerals, including iron, phytoestrogens, fibre and oils, and are used for circulation, hormone balancing and diabetes.

LADOO/LADDU

Dry-roast 2 cups of black sesame seeds and 1 cup of nuts (almonds and/or cashews) in a heavy frying pan, stirring continually for a few minutes to colour them a little, until their aroma is released. Blend them into a rough powder together with the ground seeds of 4 cardamom pods. Purée 4 or more dates to taste, with around 100g melted butter, ghee or cold-pressed oil, and mix this into the nuts and seeds. Using your hands, mould the mixture into balls, then allow to cool completely. Makes about 15.

VARIATIONS: You can vary this with different spices and add oatmeal to make the mixture more substantial.

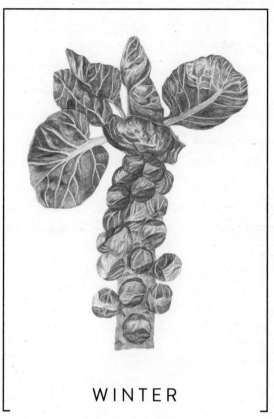

WINTER

(December–February)

BARREN WINTER APPEARS AS A SLEEPING season. Growth is halted as nature waits for light and warmth. Plants and hibernating animals are quietly feeding off their stored energy, but hidden activity is going on within plants and the soil. Rebirth is coming as the days start to lengthen after the winter solstice in December.

We, too, are preparing, but for celebration and feasting as we see out the old year, then to recover and renew ourselves for another new year. We can use the dark days of January for reflection and planning. We naturally turn inwards to build our own resources, staying home more often and sleeping more. A new year can motivate us to focus on our health. For the Romans, February was the month of purification, while the Anglo Saxons knew it as kale month – kale being one of the few fresh vegetables available in winter and, like all the cabbage family, a plant that helps the liver to clear the excesses of winter feasting.

Low winter temperatures affect the life cycle of plants. The cold kills many insects and pathogens that thrive in warmth and go on to damage plants. Plants can lower their freezing point to continue to use water when temperatures dip below zero, but usually only the top layer of soil freezes, the heat below enabling roots, and the soil microorganisms that exchange nutrients with them, to stay alive and repair prior to new growth. This plant dormancy in the cold months allows us to prune and transplant without harming the plant. Moreover, sufficient chilling time helps a fruit tree to generate stronger and more numerous buds.

EMBRACE THE COLD

Fundamental to being human is our flexibility in adapting to the environment, both emotionally and physically. While warm weather clearly benefits us, there are also advantages to the cold. Brief exposure to cold or cold water releases beneficial endorphins and noradrenaline. Colder temperatures can:

- Improve sleep – at night, our core body temperature drops by 2–3°C to initiate sleep. We fall asleep more quickly, sleep more deeply and wake less when it's cooler; 16°C is best for the bedroom, along with an open window for fresh air.

- Boost metabolism – being in temperatures closer to those outdoors benefits our health and helps us lose weight. Feeling chilly and shivering for just a short time every day raises our metabolic rate to burn more calories and brown fat (a special type of fat that makes heat).

- Help clearer thinking and lift mood – 2–3-minute cold showers may reduce depression.

- Boost energy levels in winter swimmers and reduce post-exercise fatigue.

- Activate the immune system – daily brief cold water stress can increase our white blood cells and other immune cells to support us against viruses and cancer.

- Ease pain – cold exposure, such as wild swimming or 30-second cold showers, triples the release of the nerve chemical noradrenaline, which reduces pain.

- Calm inflammation – exercising in the cold reduces the inflammatory response, meaning you can exercise for longer; race times for running in the cold are even faster.

- Improve circulation – blood vessels are muscular, so give them a workout by alternating between hot and cold to tone them and so improve their response to high blood pressure and stress in general.

You can benefit, too, by taking a 2–3-minute cold shower, gradually lowering the temperature to 20°C, or have a short shiver every day.

While cooler weather is linked to longevity in animals, it's not yet proven if it does the same for people. But it's a good idea to adjust your home thermostat to reflect a little closer the temperature outside, put on more clothes to keep warm, saving energy and saving you money, to boot – a shiver each day keeps the doctor away!

WINTER SUN AND SNOW

During the short winter days, getting what little sunshine we have is key, especially in the morning. Exposure to the blue light of day synchronizes our many body clocks to maintain our circadian rhythm and keep us feeling well and resilient to infections. If it snows, make the most of this brief, wondrous moment to play in or crunch a path through it. Snow can reflect an extra 80 per cent of light, which almost doubles our exposure to UV rays. While this may improve our mood and raise vitamin D levels, we need to protect our skin from burning more quickly, particularly at higher altitudes, so it is wise to use sunscreen. However, as our eyes need the light to maintain our circadian rhythm, use sunglasses only when they are really needed in the acute glare of snow or strong sun.

OPPORTUNITIES OF THE SEASON

Like plants storing up nutrients in their roots, we need those same roots and similar energy-rich foods to stand up to the cold weather. Although plant foods are more scarce, this is a good season for wild game, shellfish, winter cheeses and truffles – all of which are nutritionally rich.

SEASONAL FOODS

Jerusalem artichokes, turnips, celeriac and parsnips are among the roots in season. And when it's cold outside, we crave warming stews and soups, so beans, mushrooms, nuts, seeds and spices are nourishing foods to include in them. Traditional Chinese medicine advises two-thirds cooked food to one-third raw in the winter, and vice versa in the summer.

Some cultivated foods are hardy enough to withstand the cold and snow – leeks, spinach, chard and cabbages (including kale), turnips and parsnips. Add any of the following ingredients to your cooking or to make teas to ward off winter ills: all herbs, whether fresh, dried or frozen; warming circulatory spices, such as cinnamon, ginger, chilli, black pepper and horseradish; lung decongestants, such as mint, hyssop, rosemary, thyme and garlic; and the antiviral elderberry, rosehip, bilberry, blackcurrant or cranberry.

WHAT'S IN SEASON

PLANT FAMILY	EXAMPLES
Beet *Amaranthaceae*	Beetroot, perpetual spinach, chard Growing wild – fat hen/goosefoot, orach, sea beet, sea purslane
Onion *Amaryllidaceae*	Leeks
Carrot *Apiaceae*	Celeriac, parsnips, coriander Growing wild – hogweed seeds (*Heracleum sphondylium*), alexanders (*Smyrnium olusatrum*), ground elder, cow parsley/wild chervil, wild fennel **(caution: many wild species are toxic, so double-check identity)**

Daisy *Asteraceae*	Parsnips, Jerusalem artichokes, salsify (*Tragopogon porrifolius*), scorzonera or black salsify (*Scorzonera hispanica*) Growing wild – dandelion (root and leaves), burdock root, sow-thistle
Birch *Betulaceae*	Hazelnuts and cobnuts (*Corylus avellana*)
Cabbage *Brassicaceae*	Kohlrabi, horseradish, winter cabbage, Brussels sprouts, kale, cauliflower, swede/turnips, pak choi, mustard leaves Growing wild/herb – watercress, winter/land cress, bitter cress, shepherd's purse
Chickweed *Caryophyllaceae*	Growing wild – chickweed
Cucumber *Cucurbitaceae*	Pumpkin, squash
Oak/beech *Fagaceae*	Sweet chestnuts (*Castanea sativa*)
Walnut *Juglandaceae*	Walnuts (*Juglans regia*), hickory (*Carya* species)

Mint *Lamiaceae*	Mint (indoors), rosemary, winter savory, thyme, lavender, marjoram, oregano
Bay *Lauraceae*	Bay/bay laurel
Rhubarb *Polygonaceae*	Rhubarb Growing wild – sorrel
Rose *Rosaceae*	Apples, pears Growing wild – rosehips, hawthorn berries/haws, crab apple
Citrus *Rutaceae*	Oranges, Seville oranges, blood oranges, tangerines, satsumas, mandarins, lemons, limes
Nettle *Urticaceae*	Growing wild – nettles
Mushrooms *Boletaceae, Agaricaceae, Physalacriaceae, Marasmiaceae, Pleurotaceae, Cantharellaceae* and *Meripilaceae*	Mushrooms, wild and cultivated: cep/porcini (*Boletus edulis*), chanterelle and girolle (*Cantharellus cibarius*), enoki (*Flammulina velutipes*), chestnut/portabella (mature button or common mushroom; *Agaricus bisporus*), shiitake (*Lentinula edodes*), maitake (*Grifola frondosa*), oyster (*Pleurotus ostreatus*), horn of plenty/black trumpet (*Craterellus cornucopioides*)

Truffles *Tuberaceae*	Wild black or brown truffles (*Tuber uncinatum, T. aestivum*), black Périgord (*T. melanosporum*), white, northern Italian (*T. magnatum*)

WHAT TO PICK AND EAT IN WINTER

The winter landscape is not as empty of food as it may at first appear. Wild cabbage family plants such as watercress, land cress or American cress (*Barbarea verna*) and winter cress (*B. vulgaris*) are remarkably tough and thrive despite snow and frost. They are high in vitamin C, as well as giving valuable liver support. Look for the tiny, bright-green leaves of chickweed, often found at the foot of trees, used to reduce eczema itching topically and internally. The roots of the dandelion, with its antiviral and digestive properties, and the liver-supporting greater and lesser burdock (*Arctium lappa* and *A. minus*), are especially good to eat, as their sugars are concentrated in the roots now that flowering is over. Alexanders (*Smyrnium olusatrum*), from the carrot family, which begins new leafy growth in the autumn, offers leaves and the unique flavour of its black seeds atop tall stalks. These plants are found near coastal areas, on roadsides and at the edges of woodland. The peppery seeds with a hint of myrrh, which may account for its name *Smyrnium*, were used for centuries to spice up foods before imported black peppercorns became more affordable. Berries and hips, especially rosehips, found all over the UK, and orange sea-buckthorn berries, mostly from coastal areas, can still be found. Rosehips become sweeter and softer after a frost; we can reproduce this 'bletting' by freezing and then thawing them.

Truffles

These underground fungi, known as 'black gold', are highly prized for their pungent, musky and umami tastes. Truffles are a product of the remarkable partnership between the fungus and the roots of oak, beech, birch and hazel trees, among others. They were hunted mainly with dogs to sniff them out, but the UK's once thriving truffle industry disappeared after a prohibitive dog tax and World War I. Modern farming contributed further to the loss by removing woodlands, hedgerows and the biodiversity truffles need to grow.

If you are lucky enough to find them, or buy them, all that's needed is to shave or grate a small amount of truffle over pasta, risotto or vegetables. Like mushrooms, truffles contain many health benefits. As well as being 20–30% protein, truffles:

- Host a complex microbial community that benefits our own gut microbes

- Contain potent antimicrobial aromatic or essential oil compounds

- Protect the liver

- Have cholesterol- and sugar-lowering properties

- Are anti-inflammatory

- Support the immune system

Tip: For an insight into the relationships between people's land, food and culture, visit food festivals in the UK, Europe and beyond. Villages display and celebrate their beloved local produce – truffles, olives, grapes and wine or chestnuts. You learn how they grow and eat these local foods and, equally important, the role the food plays in their culture.

Citrus fruits

The most medicinal parts of citrus fruit, as in other fruits and vegetables, are their skin, pith and pips – wherever possible, use the entire fruit. We usually throw away the pith or skin of an orange or satsuma, but these colourful layers have twice the vitamin C of the juice, along with

anti-cancer and other health-giving chemicals. The
following benefits are for all fruits in the citrus family –
lemons, sweet oranges, bergamot, neroli, mandarins,
satsumas and tangerines.

- Peel – just take a small piece, say the size of
  a 50p coin, and add it to a smoothie to mask
  the bitterness. However, a little bitterness is
  worth getting used to, as these polyphenol
  flavonoids have antibacterial, anti-cancer and
  many other protective qualities. They also
  keep blood vessels supple, to protect us against
  inflammation and high blood pressure. The
  citrus skin's essential oil, including limonene,
  stimulates liver detoxification, tissue healing
  and circulation. Its fibre feeds our gut
  microbes and encourages regular bowel
  function. Grate the peel to add its zest to
  all kinds of foods – soups, salads and sauces.
  Dried citrus peel works well in teas and
  looks pretty.

- Pith – the spongy white layer just beneath the
  peel – is one of the highest sources of pectin,
  the soluble fibre that binds cholesterol, feeds
  our gut microbes and aids regular bowel
  movements. Pectin is the gelling agent that
  sets jams and marmalade.

- Flesh – while the juice contains vitamin C and some flavonoids, its high level of fructose, or fruit sugar, quickly raises our blood sugar, so it's much better to eat the flesh and seeds with their fibre, rather than drink the juice alone.

- Seeds – the seeds are antimicrobial and antifungal, inhibiting common gut and skin infections.

Marmalade is made with the highly seasonal Seville orange (*Citrus aurantium*), which is available only for a precious 6 weeks from the end of December. Although very high in sugar, it's another way of including the whole fruit, with its medicinal skin, pith and seeds. Also, how we make and eat it can modify the sugar hit and applies to all jams.

- Make a low-sugar version, using half the sugar and more lemon juice for pectin, which makes it set – it works well and still tastes sweet, but may not keep as long.

- Use real dairy butter or nut butter on the bread for their beneficial fat.

- Choose a chewy, heavy bread made with wholegrain wheat or rye, or a sourdough, as the fibre slows down the release of sugar into the bloodstream.

- Include an egg or other protein (nuts, beans, fish, meat) at breakfast with your toast and marmalade, as this will slow the otherwise too fast rise in blood sugar.

ORANGE AND APPLE PUDDING

Celebrate oranges in this deliciously tangy yet sweet pudding, with its digestive and immune-stimulating spices.

4 apples
3 large oranges: grated zest and juice
¼ teaspoon ground cloves
1 teaspoon ground cinnamon
3 tablespoons honey
50g ground hazelnuts, cobnuts or almonds (best with their skins)
20g mixed roughly chopped nuts (any two of almonds, hazelnuts, walnuts, cashews, Brazils, pine nuts and pistachios)

Preheat the oven to 250°C/500°F/gas mark 9.

Cut the apples into 1cm slices, complete with the skin, core and pips, and layer them in a shallow baking dish (if you like, discard the core when eating). Mix the orange juice, zest, spices and honey and pour over the fruit. Sprinkle over the ground and chopped nuts, cover with foil and bake for 45 minutes. Serve with yoghurt, crème fraîche or cream. Serves 4–5.

VARIATIONS: Experiment with other spices – cardamom, sliced fresh ginger, star anise and nutmeg all act as digestives and antiseptics. If you have any rosehips, add those for an extra hit of vitamin C, flavonoids and colour.

Wild game

Deer, rabbit, grouse, pheasant, duck and wood pigeon become available through autumn and winter and can be some of the healthiest meats to eat. We reap the benefits of everything the animal has eaten – albeit transformed. A wild animal has lived a natural life, eating diverse plants and smaller animals and insects, themselves rich in minerals, phytonutrients and omega-3 oils. This natural variety and lifestyle make the wild animal healthy, which, in turn, boosts the health of those who are fortunate enough to eat it.

Eating nose to tail is the animal equivalent of eating root to stem. Both approaches respect and understand nature by making use of all its parts and avoiding waste. So keep the bones for stocks for soups and risottos, and the livers for pâté.

Winter cheeses

Cheese is a fermented food, so it has multiple health benefits from its microbes and the valuable volatile oils, amino acids, proteins and lactic acid these produce. Cheese made from animals which have eaten many different plants and grasses, and from unpasteurized milk with its more diverse community of microbes, will be more complex in its taste and in its health benefits.

Winter is an exciting time in the cheese world. Although cows are milked year round, cheese made with cow's milk varies in flavour depending on the quality of the pasture the cows have been grazing on. A cheese made when grass is new and lush will taste different from a cheese made when the grass

is less rich at the end of summer. In winter, a diet of fermented grass, known as silage, creates milk high in butterfat and protein, which makes for especially creamy, musky cheeses such as Époisses or the goat's cheese Chevrotin des Aravis. One of the most eagerly awaited winter cheeses is Vacherin Mont d'Or, strapped in its wooden box. Made only in autumn and winter, from the milk of Alpine cows fed exclusively on high mountain grass and hay with wild flowers, such as deep-blue gentians, it has a sweet yet tangy taste. Bake it in its box until bubbling and serve with potatoes and other vegetables.

CHALLENGES
OF THE SEASON

Viral epidemics such as influenza and norovirus peak in December and January and may persist into early spring. Fortunately, there's much we can do to strengthen our immunity to these viruses and lessen their effect on us. The long, dark days can sometimes affect our mood, and depression and seasonal affective disorder (SAD) are prevalent in the winter months. SAD affects one in three of the UK population. Again, nature comes to the rescue and offers remedies to raise our spirits and restore our balance.

IMMUNE-BOOSTING SOUP

This truly pungent, medicinal soup is packed with immune stimulants, so make up a flaskful to warm you up on a walk and prepare for the powerful tastes needed to manage winter viruses.

7–10cm fresh ginger, grated

1–2 medium chillies, finely sliced, with seeds

2 large onions, peeled and sliced

12 shiitake or other mushrooms, thinly sliced

2 tablespoons dried *Echinacea angustifolia/ purpurea* root, tied up in a small muslin cloth

5 tablespoons fresh or frozen, or 2 tablespoons dried, of any one of the following berries: elderberries, rosehips, bilberries, cranberries, blackcurrants or other black berries

2 litres vegetable or chicken stock

8 crushed garlic cloves

1 tablespoon chopped fresh thyme

2 tablespoons sweet miso paste

Sauté the ginger, chillies and onions in a pan until soft. Add the mushrooms and fry lightly for 5 minutes. Add the *Echinacea* root bag along with the berries and stock, and simmer for 20 to 30 minutes.

Before serving, remove the muslin bag and stir in the garlic, thyme and miso. Serve straight away, or put into a vacuum flask for later. Serves 4–5.

NOROVIRUS, THE WINTER VOMITING BUG

Norovirus is a very contagious virus that causes abdominal pain, nausea, vomiting and diarrhoea, usually lasting for between 1 and 3 days. We remain contagious for up to 3 days after our symptoms pass, and the virus itself can survive for even longer. It's spread by lack of hand washing after going to the toilet, so good hygiene is key to avoiding norovirus. Washing your hands thoroughly before preparing food and washing all raw foods before consuming them will help to lessen the risk of infection. If you can, stay away from anyone infected.

Protect yourself against norovirus

In addition to the hygiene advice above, foods offer plenty of phytochemicals to support immunity and prevent norovirus taking hold of your body.

- Eat naturally antiviral foods, such as grape seeds, elderberries, oregano, pomegranates, mulberries, cranberries and their juice, citrus fruits and green tea.

- Make a litre of herb tea with at least 2 or more of these herbs: yarrow, blackberry leaves, oregano, thyme, chamomile, meadowsweet and cinnamon – infuse about 30–40g of dried herbs along with 4 or 5 thin slices of fresh ginger (or ½–1 teaspoon of ground ginger), with, if you like, ½ teaspoon of dried or 1 teaspoon of fresh chopped orange or other citrus peel. Leave to infuse till warm/cool, then drink throughout the day and take the final cup to bed.

- Enjoy the immune-boosting soup on page 168 and take some regularly, or include immune-boosting foods – mushrooms, cinnamon, thyme, rosemary, dandelion leaves, garlic and onions – in your daily menu.

- Keep hydrated with warming herb teas (see page 170), and add seaweed, nettles and other hydrating foods to smoothies and other drinks.

Antiviral spray

Alcohol hand rubs don't kill norovirus, so instead make up this alcohol-based spray by mixing 25 drops of oregano (*Origanum vulgare*) essential oil with 1 teaspoon of vodka (as a dispersant) and 3 tablespoons of water, or just 50ml of water, in a glass spray bottle. Shake well to disperse before using.

Alternatively, use a strong green tea as a spray, as it contains antiviral catechins (a type of tannin that gives tea its astringent taste, and present in many other foods). Spray both your face and your hands and, if not too indiscreet, the area around you as a humidifier to help break up virus particles. In fact, you can even spray it into your throat to boost its immune response to infection.

FLU

Influenza or flu epidemics are tied to cold, dry air and so tend to peak in the winter months. When infected with the flu virus, we breathe out virus particles into the air around us, as well as sneeze and cough them out. Protect yourself from infection with an antiviral spray (as for norovirus, page 171) or the air antiseptic spray (see page 122), as they also humidify the surrounding air.

Flu, unlike a cold, comes on suddenly, with body aches, sore throat and a cough, loss of appetite and above all exhaustion – our first instinct is to go to bed. Sleep is the best thing to do; it has a powerful effect on the immune system – the body is trying to sleep itself well. The virus forces us to slow down completely, take very little food and just rest to recover, which can take some time.

A flu tea

Herbal teas truly come into their own when dealing with flu – they are comforting and gentle, and have just the right properties for recovery. Herbs such as elderflower, lime blossom (*Tilia cordata* or *T. europaea*, the linden tree), peppermint and yarrow encourage sweating by opening pores to release heat in a high fever. Make 1 litre of tea with 4–5 tablespoons of herbs (say 2 tablespoons each of dried lime blossom and elderflower) and include some grated or sliced fresh ginger. Infuse for at least 20 minutes before drinking for a good, strong mix. Vary the herbs with peppermint and yarrow from day to day. You can also add green tea (*Camellia sinensis*), which is strongly antiviral. Ginger is warming, while dispersing heat to our skin, and quells the nausea that can accompany flu.

Post-flu nourisher

If you're feeling too weak to do much cooking, try this easy soup as a tonic. If you have any nettles – fresh, frozen or dried – add these at the start for extra medicine.

Put a head of garlic and 12 whole mushrooms into a pan with about 750ml of stock or water. Add several sprigs of thyme (or rosemary), 5 tablespoons fresh or frozen (or 2 tablespoons dried) of one these berries: elderberries, rosehips, bilberries, cranberries, blackcurrants or other black berries, 1 tablespoon of pot barley or lentils, 1–2 chopped chillies, and a few slices of fresh ginger. Cover and simmer for 20 to 30 minutes. Off the heat, stir in a freshly crushed garlic clove, some more chopped thyme and 2 dessertspoons of sweet miso paste, and savour the warmth.

SEASONAL AFFECTIVE DISORDER (SAD)

This seasonal depression – often called 'winter blues' –
tends to have more severe symptoms in the winter
months, starting as the days get shorter in the autumn.
Disrupted circadian rhythms are among the causes. It
seems that some people's body clocks fail to adjust to
the shorter daylight hours.

Getting as much morning and afternoon light
as possible, along with outdoor exercise, will help
reboot our circadian rhythm, make us feel better and
strengthen immunity. Avoid sunglasses if you don't
really need them, especially in autumn and winter, as
they prevent light from reaching the eyes, which helps
to regulate normal circadian rhythm.

Rebooting circadian rhythms

Sleep is often affected in SAD, and improving this and
eating supportive foods help to reset our body clock.

- To encourage sleep, make yourself a hot
  valerian chocolate: this comforting drink
  with feel-good elements in the chocolate will
  balance the tiny amount of caffeine unless you
  are very sensitive to it, and the valerian has a
  sedative quality. Simmer 1 teaspoon of
  chopped valerian root in 1 cup of oat milk,

with a square of 70% chocolate or
1 teaspoon of cocoa powder and a
little honey. Pour into a mug, take
it to bed and drink with your
awareness fully on every
comforting sip, then put the
light out; you could even try
drinking it in the dark.

- Do the same mindful relaxation (as with
  the hot chocolate, above) with a herb tea, using
  one or a mix of chamomile, lime blossom,
  lemon balm, passionflower, hops, rose petals
  or a little lavender.

- Kiss someone before you sleep. Kissing
  decreases the stress hormone cortisol and
  increases serotonin levels in the brain to raise
  mood, which makes falling asleep easier.

- Try the 4–7–8 slow-breathing technique: count
  4 while breathing in, hold for a count of 7 and
  breathe out for 8. Such breathing slows down
  the heart rate to bring the body into a resting
  state and relieve stress or tension.

- Relaxing in a warm bath before bed will
  improve sleep. The water should be about
  39°C/102°F and you need to lie back and relax
  for 30 minutes in the early evening.

Mood foods

A key way to supply our bodies with the feel-good chemicals they need is by feeding our gut microbes with plenty of fibre. They are our all-important partners that digest the fibre to make essential compounds, such as short-chain fatty acids, and promote neurotransmitters such as serotonin and dopamine. This is another good reason for eating the skin, pith and seeds of vegetables and fruit. Along with this, our brain and nerves (which are at least 60% fat) also need foods with essential fatty acids – omega-3 and -6, found in oily fish, avocados, nuts, seeds, especially linseeds, and green leafy vegetables. Zinc-rich foods – shellfish, meat, pumpkin seeds – are important, too, as zinc is essential for nerve cell communication.

SEASONAL GROWING AT HOME

Some vegetables, such as shallots and garlic, can be planted into the ground even in December. You could also sow broad beans, but these will need protection from the cold and from hungry birds, mice and squirrels, so cover with a cloche. If you planted snowdrop bulbs (*Galanthus*) after flowering while still 'in the green' back in the spring, the flowering of their little white heads will gladden the heart.

THINK MICROGREENS

There's no need to buy bagged salad leaves when you can grow your own, including micro versions, all year round, though they'll need to live inside on a windowsill or in a greenhouse through the winter. This is the same principle as sprouting beans (see page 110) but uses much smaller seeds.

- Mint can be moved inside in its pot for a never-ending supply of its fragrant leaves.

- Microgreens – the seedlings of herbs or leafy vegetables – both look pretty and provide a burst of nutrient-packed flavour. Basil,

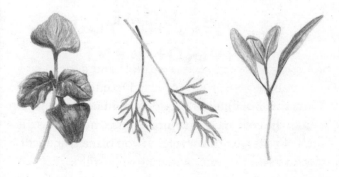

beetroot, broccoli, coriander, fennel, mizuna, mustard, radish, rocket, spinach – most take a mere 7 to 10 days to grow before you cut them.

INDOOR COMPOSTING

If outdoor space is minimal or missing altogether, it's possible to compost food waste indoors. Follow the urban Japanese example of bokashi bins (*bokashi* means fermentation). Simply turn kitchen scraps into compost by putting them into a bin with a tight-fitting lid, sprinkling with 'bokashi bran' (it's inoculated with active bacteria that do the hard work), and 2 weeks later you're left with decomposed matter that you can treat like compost and dig into the garden, if you have one, or mix into a windowbox or planter.

BEAUTIFYING YOUR ENVIRONMENT

Winter is about indoor plants. They all help to purify our air and some also permeate it with scent, such as the beloved Christmas tree, which is another medicine in itself.

PLANTS TO PURIFY YOUR HOME

As well as lifting our spirits, plants help clean the air we breathe. They are especially helpful in the winter, when windows tend to be closed and there is less circulating air. They not only clean our air, but produce oxygen and encourage moisture. Pollution may be more visible in the summer months, especially in cities and big towns, but have you ever considered the idea of indoor pollution? Indoor air can be more harmful to health than that outside, and that's where plants can help. All plants contribute to cleaning air, but those listed below are particularly effective purifiers.

- Aloe (*Aloe vera*)

- India rubber tree, rubber plant (*Ficus elastica*)

- English ivy, common ivy (*Hedera helix*)

- Boston fern (*Nephrolepis exaltata* 'Bostoniensis')

- Mother-in-law's tongue (*Sansevieria trifasciata*)

- Nephthytis (*Syngonium podophyllum*)

- Fern arum (*Zamioculcas zamiifolia*)

- Bamboo palm, lady palm (*Rhapis excelsa*)
- Peace lily (*Spathiphyllum* species)
- Spanish moss airplant (*Tillandsia* species)

ALOE VERA

If you only have space for one plant, opt for *Aloe vera* – its gel helps to heal burns and skin wounds as well as calm digestive complaints:

- For burns on unbroken skin, split open a bit of leaf, put the gel side on to the burn and wrap round with some clingfilm or Micropore tape.
- For cuts, grazes, eczema and psoriasis, just apply the gel twice daily.
- For constipation, make a smoothie by blending a dessertspoon of whole aloe leaf with a glass of water, a thick slice of cucumber, lemon juice and a few leaves of spinach/

lettuce or other green leaf. Even more effective, soak a dessertspoon of linseed in a glass of water for an hour until it has thickened as the seed absorbs the water and add this to the smoothie. You can also use the linseed soaked in water alone as a constipation remedy.

Keep the plant growing, with only a little water, and it will make 'babies' that you can transplant into new pots and give to friends.

CHRISTMAS TREES

Choosing a tree for Christmas and health means only one thing – go for the scent and a real tree, not plastic. You reap the rewards of forest bathing – *shinrin-yoku* – in your own home, breathing in the beautiful, aromatic oil of pine leaves to clear your head and your lungs. Those with the best fragrance are the traditional Norway spruce (buy with its roots, as it drops its leaves fast), Scots pine and Douglas and Fraser firs. If you can, plant it in a garden afterwards. Try to find a tree that's been grown sustainably as close to your home as possible – you can even go and dig one up in places where you are allowed to. You can also rent a tree, which is delivered to your home every year and replanted as part of the deal. Whichever variety of tree you choose, do recycle or replant your tree afterwards. There is huge waste every Christmas – only 10% of the 6 million trees in the UK are recycled for composting and wood chipping; sadly, the rest goes into landfill. Recycling this valuable tree as garden waste at your local centre will provide nutrients for depleted soil and nurture future plants and wildlife.

- What's more, you can make a delicious and decongestant tea out of a few of your Christmas tree's needles. Just strip a teaspoon of needles from a branch, cover with boiling water, and simmer for about 15 minutes.

- If you plant your tree, then be sure in the spring to pick a few little green shoots to add to salads – they are soft and lemony, and the pinene and other aromatic compounds protect our immune systems. It's just what our hunter-gatherer ancestors did as they roamed through forests, nibbling shoots for food and medicine.

A BURST OF COLOUR AND SCENT

You may be familiar with the name witch-hazel, as its twigs are distilled to make witch-hazel water, an astringent and antiseptic. The species native to North America – *Hamamelis virginiana* – has long been used for cuts and bruises, acne, eye puffiness and as a skin cleanser and toner. The shrub flowers in mid to late winter, with spidery blooms of yellow, red and orange, and is fragrant, almost spicy. Put a cutting in a vase to bring its winter scent into your home.

By the end of January, snowdrops will be pushing up through frost or snow to herald spring. If you planted them in the garden earlier (see page 30), now comes the reward of picking a few of these delicate white flowers to put into a vase. The shrub daphne (*Daphne odora*, *D. mezereum*) also flowers early in the new year and has the most gorgeous and powerful lemon scent – a sprig by the bed lifts your spirits and so may help you fall asleep.

FLOATING FLOWERS

Winter-flowering hellebores, the Christmas rose *Helleborus* and its hybrids, are winter beauties in the great outdoors that often aren't enjoyed to the full, as

their exquisite flowers face downwards. Bring their beauty indoors by floating some flowers, face up, in a small, shallow bowl of water; simply pinch off a few flower heads, lay them close together on the surface of the water and delight in their pattern and colour.

WELL-BEING, MOVEMENT AND SOCIAL CONNECTION

While December may be full of celebration and social gatherings, winter can also be a reflective time. Dark nights may lend themselves to reading inside, but take advantage of crisp, clear nights to head out and wonder at the moon and stars.

BECOMING MINDFUL

A 3-minute body scan daily: sit in a chair, feet flat on the floor, hands relaxed. Close your eyes and breathe gently. Notice how your feet feel – warm, cold, any pain? Moving up, focus for a few seconds on each area – legs, hips, bottom, belly, chest, arms, hands, head and face – any discomfort, hunger, stress, emotion? The body will tell us what it needs if we listen. And it reduces stress, calms our busy minds, and helps us concentrate and pay attention to the present.

SUPPORT A FARMERS' MARKET

Buying locally produced food is one of the most effective ways to support the planet. It's also a chance to reconnect with nature through food. Talking to food producers gives a fascinating insight into what is involved before food reaches our plate. Markets usually give you the choice to buy organic or pasture-grown foods, where creating healthy, complex soil is fundamental to farming. Essentially, the quality of food depends mainly on the soil that grew the plants or on the fodder and husbandry of the animals we eat. Quality tastes good – the more taste buds tickled, the more complex our food and high in phyto-nutrients. We pay for poor-quality food with poor health.

The farmers and growers tend to be smaller-scale and take longer to grow their plants or animals. So buy your own food-as-medicine through your local farmers' market. It's also one of those too-rare public places where you can meet people and talk about food – many conversations happen at stalls about how to cook and use the produce. And you'll likely save on food miles, food energy and food waste.

READING AND READING ALOUD

Something easy and rewarding is to read aloud in a group, to children or simply to your partner. Being read to doesn't have to be the sole preserve of childhood, it's for everyone. Introducing this to others who have grown up without storytelling or poetry is a wonderful way of re-experiencing the pleasure and passing it on. Shared reading also engenders a human support system that improves our well-being and builds stronger communities. It reveals that as individuals, we have much more in common than we think. It can also be life-changing, especially for children and for those of us who are isolated, with pain or dementia, not to mention fun. The benefits are widely recognized, in fact, as you can become a 'reader in residence' in a library, a GP surgery, a care home, a prison or even a supermarket.

BEING AMONG WILDLIFE

Even in winter, there are activities that develop our skills and our relationships with nature. We learn to see the season through the eyes of the animals or birds that live it. Being outdoors with wildlife helps us slow down to the pace of the natural world.

- Tracking animals – enter their way of being and their landscape. We shift awareness from our head to our body to use our senses to their fullest. Most of us now use only a fraction of our senses, as we no longer hunt for food. Following an animal means stretching our sight, smell, hearing, touch and, above all, our imaginative intelligence. A footprint of any kind – animal or human – leaves clues about speed, direction and physical and mental condition. Learning with an expert how to read these signs can be deeply satisfying, and paying such close attention can foster patience and an inner calm.

- Enjoying the birdlife – thousands of migratory birds come to the UK in the winter, and bird sanctuaries and nature reserves offer the perfect places to view them. But a humble bird feeder in a garden or on a balcony will give food to your local birds and pleasure to you when they come to visit.

- Learning woodland husbandry – from carving and working with wood to charcoal-making,

flint-knapping, and 'social forestry', there are
activities galore to care for the woodland
and our own well-being; even stacking wood
can be done mindfully and with beauty.

- Identifying winter trees – look for small
details, such as the pattern and hairiness
of hazel shoots or the shapes of pine needles
to identify different species of Christmas
tree

AWE, NATURE AND THE UNIVERSE

Gazing up at the moon and stars on a dark, winter
night can inspire awe and wonder. These feelings are
actually health-giving. Just feeling the sheer scale of
the universe and the beauty of the moon – even
standing amid tall trees – can lead us to be more
altruistic, less entitled, more aware of the strengths of
others and less stressed by the challenges of daily living. These brief experiences give us a sense of being
part of a larger social collective and are good for the
immune system. Feelings of awe awaken our imagination and can often lead to an epiphany or a striking
realization about life. For more practical fun, phone
apps, some free, offer immediate interpretations of
the sky at night.

CELEBRATIONS, ANNIVERSARIES AND SPECIAL DAYS

The months of December to February are full of celebrations; perhaps we need these most at this time of year, with such long nights. There are many ways to be with and give to others, preparing events and making food for family, friends or people who are homeless. The use of spices at Christmas in food and drink is timely, given they improve our resilience to infection and warm us up, stimulating digestion and circulation.

WINTER SOLSTICE AND CHRISTMAS DAY

The winter solstice (21 December) is the shortest day of the year in the northern hemisphere, when the North Pole is tilted furthest away from the sun. From now on the days begin to lengthen, and many cultures consider it a time of hope. Pagans know it as Yule and celebrate the return of light and the birth of the new solar year at sunrise at ancient sites, such as Stonehenge in the UK and Newgrange in Ireland. Traditional celebrations in Japan involve bathing with yuzu, a citrus fruit high in vitamin C and circulatory properties to guard against colds, treat roughness of skin, warm the body and relax the mind. Christmas was originally set at the winter solstice by the Romans, but the date slipped to 25 December with calendar changes.

At Christmas, make these health-giving recipes or give them as presents to friends – za'atar, a hot and spicy mull of hibiscus, and a colourful digestive tea with orange peel all help to counter any overeating.

ZA'ATAR

Za'atar is an Arabic word referring to herbs of the mint family, particularly oregano and thyme. It's the same word that's used to describe the much-loved and endlessly varied fragrant flavouring made of dried and ground herbs, sesame seeds, sumac and salt. As a traditional food of the Levant region, preparing za'atar is a way of remembering the Middle Eastern origin of Jesus and Christianity – and eating it boosts health. Sumac – the citrus-tasting red fruit of the staghorn tree and other *Rhus* species – is a powerful antioxidant that may lower blood pressure, while thyme and oregano are immune-stimulating and antiseptic.

Make your own za'atar by grinding 1 part lightly toasted sesame seeds, 1 part dried thyme, 1 part dried oregano, 1 part sumac, with sea salt to taste. Store in a jar in the fridge for up to a week.

Dip bread into olive oil and then into a little za'atar, or sprinkle it on to meat, salads or vegetables.

FESTIVE MULLED HIBISCUS

This tangy and spicy drink tastes just like mulled wine without the alcohol, but you can easily add spirit or wine if you like. Increase the spices for

a stronger mull and even more digestive support. Hibiscus lowers blood pressure and cholesterol and is a good source of vitamin C.

Crush 6–8 cloves and a stick of cinnamon together. Put them into a large pan with 2cm of grated fresh ginger, some grated nutmeg, 25g of whole dried hibiscus flowers and 1 litre of water. Cover and simmer gently for at least 25 minutes, then add honey or sugar, to taste. Strain into glasses. Serves 4.

VARIATIONS: You can also add black pepper, bay leaves and allspice, along with orange juice and/or orange/lemon peel.

DIGESTIVE CITRUS TEA

Make this wonderful orange-coloured tea to stimulate digestion, deal with any Christmas excesses, and reduce bloating and wind.

Dry the peel from any orange citrus fruit slowly over a couple of hours in a very low oven. Chop into small pieces and add ½ teaspoon to a cup of herb tea, with any one or a mix of fennel, caraway, cumin and cardamom seeds, aniseed, cinnamon or ground ginger.

Pack in small cellophane bags, with raffia or ribbon and a label, and give this homemade tea to friends.

HANUKKAH

The Jewish Festival of Lights spans 8 days – the date varies every year, as it starts on the 25th day of Kislev, the month in the Jewish calendar that occurs at about the same time as December. The word 'hanukkah' refers to the rededication of the temple following a battle that won religious freedom for the Jews, and the miracle of a small jug of olive oil for lighting a lamp that should have lasted one day but continued to provide light for 8 days. The Menorah's eight candles represent the 8 days, with an extra 'helper' candle built into the candelabra. During the festival, each night one new candle is lit. The importance of oil is recognized by preparing foods such as potato latkes – fritters that can also be made with vegetables, cheeses and legumes. Grated raw potato (or beetroot, sweet potato or other root), onion, egg and cornflour or chickpea flour are combined, then small portions are fried in oil. The onions and the protein of the eggs make it a wholesome and delicious dish; it's even more so if made with purple potatoes or beetroot with their protective polyphenols.

KWANZAA

As a response to the Los Angeles riots of 1965, an African-American professor set up Kwanzaa to celebrate the 'communitarianism' of African heritage and its values. It contrasts with the increasing consumerism of Christmas. Across the African diaspora, people gather with family and friends to exchange gifts and to light a series of black, red and green candles. Each of the 7 days (26 December to 1 January) is dedicated to a value that fosters collectivity, responsibility and a love of family and the ancestors. The candles symbolize the seven basic values of African family life: *umoja*, unity of family, community, nation, race; *kujichagulia*, self-determination, taking responsibility for yourself; *ujama*, collective work to help each other; *ujamaa*, cooperation in business; *nia*, remembering and restoring African customs and history; *kuumba*, creativity to make communities better; *imani*, faith in people, families, teachers, leaders and the righteousness of the African struggle. Some people fast up to the feasting on the sixth day, when everyone gets together to enjoy such typical dishes as black-eyed beans with peppers, peanut soup, jollof rice, okra and greens, yam and coconut pie.

IMBOLC

Imbolc falls midway between the winter solstice and the spring equinox – on 1 February – when the Celts celebrated Brigid, the goddess of fire, healing and fertility, to ensure that she revived the cold land with the onset of spring warmth. The lighting of fires sits at the heart of Imbolc and the promise of the sun over the coming months for a successful harvest. As sheep gave birth, precious new milk or 'olmelc' came, for making cheese and butter, and ingredients associated with the sun – egg yolks and honey – became central to Imbolc. In the Christian calendar, this holiday was renamed Candlemas, when candles are lit to remember the purification of the Virgin Mary. As in Imbolc, Candlemas featured foods such as pancakes, as their grain flour and round shape symbolized the sun and the harvest.

BARLEY PANCAKES

Unlike regular pancakes, this recipe uses barley, one of our cheapest and healthiest foods, which you can also use in risottos and the ultimate pulse-grain dish (see page 111).

Barley contains a type of gluten protein, but is usually more digestible than wheat gluten. Its beta-glucans, like those in oats, benefit the heart, blood pressure and bones, and its fibre content helps maintain a healthy weight. Barley is unusual among grains in containing selenium, which plays a key role in the liver and immune system and reduces inflammation. Its fibre and lignans – protective phytoestrogens present in grains and seeds – may help lower excess oestrogen and ease menstrual and hormonal problems. Whole grain and pot barley retain more of the fibrous bran layer than pearl barley.

Either serve the pancakes with butter, or with fruit and yoghurt, or as a savoury dish with some sautéd onions, a young goat's or sheep's cheese and some za'atar to spice it all up.

50g barley flour (or grind wholegrain or pot barley in a blender till fine)

50g plain flour
¼ teaspoon salt
2 eggs
25g soft unsalted butter
100ml milk
100ml cold water
unsalted butter, for greasing

Sift the flours and salt into a blender, then add the remaining ingredients and whiz until smooth. Leave to stand for at least an hour and preferably refrigerate overnight. The batter may thicken, so thin it with water and milk to the consistency of single cream, otherwise the pancakes will be heavy.

Melt a knob of butter in a heavy-based frying pan and, when hot, add a ladleful of batter to make a thin coating, tipping out any unset mixture. Fry for 1 to 2 minutes, to a golden brown, then flip it over and fry on the other side until golden brown spots develop.

Add another small knob of butter before you fry each pancake. Transfer to a plate when ready. Makes 8–10 pancakes.

VALENTINE'S DAY

By the fourteenth century in Britain there was a belief that birds chose their mates on St Valentine's Day (14 February) and so should we. This idea may have been linked to the coming of spring or the Roman festival of cleansing and fertility in February. Whatever its origins, 14 February is now a thoroughly commercial day, so why not fall in with the fun and cook some pink or red food for your best beloved? Let your erotic imagination mix with your epicurean skills to create a meal with seasonal ingredients, such as salmon, langoustines, scallops and beetroot, and desserts with rhubarb, and cocktails or fizz with pomegranates.

Glossary

Alliin, disulphides: Sulphur-containing compounds in the onion family; allicin is the pungent smell released from alliin when garlic is crushed; anti-carcinogenic, regulate cholesterol, support liver metabolism, anti-microbial.

Alpha and beta-carotenes: Carotenoids that we convert into vitamin A, which protects cells from damage; important for our immunity, skin, heart and eye health.

Antioxidant: A substance that prevents the damaging effects on cells from an imbalance of oxidation by free radicals, a bit like the rusting of metal.

Citrulline: A precursor to another amino acid, arginine, found in high levels in the melon family (*Curcubitaceae*), especially watermelons, and produced naturally in our bodies; important for immunity, and for blood vessel and heart health.

Glucosinolates and indoles: Sulphur-containing compounds with the mustard taste in the cabbage family (*Brassicaceae*), including Brussels sprouts, watercress and horseradish; anti-inflammatory, support liver metabolism, protect against cancer.

Lignans: A type of polyphenol with beneficial phyto-estrogenic function in us. Widely found in plant foods, especially wholegrains, pulses and seeds and particularly high in linseeds.

Lycopene: A carotenoid, coloured red; it supports eye and cardiovascular health, anti-inflammatory, anti-carcinogenic. High in tomatoes, also in watermelon, apricots and guava fruit.

Phytochemicals/phytonutrients: A range of chemicals produced naturally by plants as secondary metabolites for defence against predators, pathogens and ultraviolet light, and for communication and other plant functions. They include polyphenols, carotenoids, phytosterols, coumarins, tannins, lignans and glucosinolate/sulphur compounds. They benefit humans (and animals) by protecting us against inflammation, controlling hormones, supporting circulation and helping us to manage disease through many other pathways.

Phytoestrogens: A term to describe several phytonutrient groups including lignans and isoflavones that have important hormone-balancing and possibly anti-cancer roles. We rely on our gut microbes to convert phytoestrogens into their active form. Highest levels are found in pulses and seeds.

Polyphenols: A group of colourful phytochemicals including flavonoids, isoflavones, resveratrol, lignans, tannins and curcuminoids; benefits are supporting blood vessels, balancing hormones, improving immune function and gut microbes, and providing antioxidant and/or anti-cancer activity.

Proanthocyanidins, or condensed tannins: A group of polyphenol flavonoids found in many flowers and plants as protective red, blue and purple pigments. Astringent and concentrated in the skin, seeds and bark, with high levels in grape skins and seeds, black and green tea, apples, cocoa beans, cinnamon, berries and pine bark. They improve capillary function and blood flow and reduce inflammation, and have antiviral, anti-cancer and anti-diabetic actions.

References and Resources

Introduction

Chatterjee, Rangan, *The 4 Pillar Plan: How to Relax, Eat, Move and Sleep Your Way to a Longer, Healthier Life*, London, Penguin, 2017.

FAO 2014: http://www.fao.org/3/a-i3685e.pdf. Promotion of underutilized food for food security and nutrition in Asia and the Pacific; FAO http://www.fao.org/3/a-i3022e.pdf. *Sustainable Diets and Biodiversity*, 2010.

Harrod Buhner, Stephen, *The Lost Language of Plants: The Ecological Importance of Plant Medicine to Life on Earth*, USA, Chelsea Green Publishing Company, 2002.

Martin, Cathie, et al., 'How can research on plants contribute to promoting human health?', *The Plant Cell*, Vol. 23 (5), May 2011, pp. 1685–99.

Montgomery, David, and Biklé, Anne, *The Hidden Half of Nature: The Microbial Roots of Life and Health*, New York, W. W. Norton & Company, 2015.

Panda, Satchidananda, *The Circadian Code: Lose Weight, Supercharge Your Energy and Sleep Well Every Night*, London, Vermilion Books, 2018.

*

FareShare: fareshare.org.uk – charity which saves food waste and addresses hunger by redistributing surplus food through other UK charities.

Love Food Hate Waste: www.lovefoodhatewaste.com – tips and recipes on how to use food and save waste.

Olio: https://olioex.com – app that connects neighbours and local shops in order to share surplus food.

The Zero Waster: thezerowaster.com/zero-waste-living-uk/ – guide to zero waste shopping in the UK.

Spring

Fresh air

Rook, Professor Graham, UCL, 'Microbes – The Old Friends Hypothesis', www.grahamrook.net.

Flandroy, L., et al., 'The impact of human activities and lifestyles on the interlinked microbiota and health of humans and of ecosystems', *Science of the Total Environment*, Vol. 627 (15), June 2018, pp. 1018–38.

Li, Qing, *Shinrin-Yoku: The Art and Science of Forest Bathing*, London, Penguin, 2018.

*

Forest therapy/*shinrin-yoku*: www.natureandforesttherapy.org/the-science.html – collated research on the benefits of forest bathing.

Vitamin D

Webb, A. R., et al., 'Colour counts: sunlight and skin type as drivers of vitamin D deficiency at UK latitudes', *Nutrients*, Vol. 10 (4), April 2018, article 457.

Holick, M. F., 'Vitamin D: evolutionary, physiological and health perspectives', *Current Drug Targets,* Vol. 12 (1), January 2011, pp. 4–18.

*

Norwegian Institute for Air Research exposure calculator for Vitamin D levels: https://fastrt.nilu.no/VitD-ez_quaRTMEDaNDMED_v2.html.

Vitamin D-rich foods: oily fish (e.g. sardines, mackerel, pilchards, salmon), cod liver, animal liver, egg yolk, dairy milk products.

Wild foods

Brown, Jo, 'Breeding the nutrition out of our food', in *New York Times*, 25 May 2013.

Robinson, Jo, *Eating on the Wild Side: The Missing Link to Optimum Health*, USA, Little, Brown & Company, 2014.

In season

Irving, Miles, *The Forager Handbook: A Guide to the Edible Plants of Britain*, London, Ebury Press, 2009.

*

Eat the seasons: www.eattheseasons.co.uk/index.php – details of current food in season.

Eat seasonably: eatseasonably.co.uk/what-to-eat-now/ calendar – calendar of seasonal foods.

Foraging and wild plant identification

Association of Foragers: www.foragers-association.co.uk – resource to find a local forager near you.

Foraging and plant identification courses and recipes with Robin Harford: www.eatweeds.co.uk.

Foraging common law guidelines: www.woodlandtrust. org.uk/visiting-woods/things-to-do/foraging/ foraging-guidelines.

Wild Food UK: www.wildfooduk.com – includes photographs for plant identification

Challenges of the season

Mirondo, R, and Barringer, S., 2016: https://www.ncbi. nlm.nih.gov/pubmed/27649517.

The pollen calendar: www.asthma.org.uk/advice/triggers/ pollen – to prepare for hayfever season.

*

Seasonal growing

Crawford, Martin, *Creating a Forest Garden: Working with Nature to Grow Edible Crops*, Dagenham, Green Books, 2010.

Liebreich, Karen, Wagner, Jutta, and Wendland, Annette, *The Family Kitchen Garden*, USA, Timber Press, 2009.

Morrow, Rosemary, *Earth Users' Guide to Permaculture*, East Meon, Permanent Publications, 2nd reprint edition, 2015.

Whitefield, Patrick, *The Earth Care Manual: A Permaculture Handbook for Britain and Other Temperate Climates*, East Meon, Permanent Publications, 2016.

*

Agroforestry: agroforestry.net – non-profit organization providing educational resources about trees.

No-dig courses: www.charlesdowding.co.uk – advice on organic and no-dig farming.

Predator control: www.greengardener.co.uk – advice on environmentally friendly products.

Seeds to buy and save: www.realseeds.co.uk – for the kitchen garden.

Heritage seed supplier: www.thomasetty.co.uk – for older varieties of vegetable/fruit seed.

Well-being, movement and social connection

Bratman, G. N., et al., 'Nature experience reduces rumination and subgenual prefrontal cortex activation', *PNAS*, Vol. 112 (28), July 2015, pp. 8567–72.

Castrén, E., and Kojima, M., 'Brain-derived neurotrophic factor in mood disorders and antidepressant treatments', *Neurobiology of Disease*, Vol. 97 (Pt B), January 2017, pp. 119–26.

Hanson, S., and Jones, A., 'Is there evidence that walking groups have health benefits? A systematic review and meta-analysis', *British Journal of Sports Medicine*, Vol. 49 (11), June 2015, pp. 710–15.

Manninen, S., et al., 'Social laughter triggers endogenous opioid release in humans', *Journal of Neuroscience*, May 2017, 0688–16.

*

Five Surprising benefits of walking: www.health.harvard.edu/staying-healthy/5-surprising-benefits-of-walking.

GoodGym: www.goodgym.org/ – community of runners that combines getting fit with doing good.

Volunteering

The Conservation Volunteers: www.tcv.org.uk/volunteering – community volunteering charity.

Marine Conservation Trust Beachwatch: www.mcsuk. org/beachwatch/– national beach cleaning and litter surveying programme.

National Council for Voluntary Organisations: www.ncvo. org.uk.

National Trust volunteering: www.nationaltrust.org.uk/ holidays/working-holidays – working holiday opportunities.

Raleigh International: raleighinternational.org – focuses on creating lasting change through youth engagement in natural resources, livelihood, water sanitation and hygiene and civil society.

Royal Society for Protection of Birds (RSPB): www.rspb. org.uk/get-involved.

Waterways Recovery Group: www.waterways.org.uk/wrg/ regional_groups/regional_groups – volunteering opportunities for the restoration of UK canals and waterways.

Worldwide Opportunities on Organic Farms and gardens: www.wwoof.org.uk.

A spring-clean for the mind

Thich Nhat Hanh, *Peace is Every Step*, London, Rider, 1991.

*

Mindfulness apps: Headspace; Calm, Insight Timer (free).

World Earth Day

www.earthday.org/earthday/ – information and resources on World Earth Day.

Calculate your own ecological footprint: www.earthday.org/take-action/footprint-calculator.

Tips to help you go green and protect the earth: www.earthday.org/earth-day-tips.

Summer

The sun

Hoel, D., et al., 'The risks and benefits of sun exposure', *Dermato Endocrinology*, Vol. 8 (1), January–December 2016.

Korać, R., and Kapil, K., 'Potential of herbs in skin protection from ultraviolet radiation', *Pharmacognosy Review*, Vol. 5 (10), July–December 2011, pp. 164–73.

Sies, H., and Stahl. W., 'Nutritional protection against skin damage from sunlight', *Annual Review of Nutrition*, Vol. 24, July 2004, pp. 173–200.

*

British Association of Dermatology: www.bad.org.uk.

Essential fatty acids and skin health: Oregon State University, lpi.oregonstate.edu/mic/health-disease/skin-health/essential-fatty-acids#summary.

Summer bitters

Lu, P., et al., 'Extraoral bitter taste receptors in health and disease,' *Journal of General Physiology*, Vol. 149 (2), February 2017, pp. 181–97.

Menella, J., et al., 'The bad taste of medicines: overview of basic research on bitter taste', *Clinical Therapeutics*, Vol. 35 (8), August 2013, pp. 1225–46.

Longer days, less sleep

Bayer, L., et al., 'Rocking synchronizes brain waves during a short nap', *Current Biology*, Vol. 21 (12), June 2011, pp. 461–2.

Food poisoning

Lillebæk, E. M. S., et al., 'Antimicrobial medium- and long-chain free fatty acids prevent PrfA-dependent activation of virulence genes in Listeria monocytogenes', *Research in Microbiology*, Vol. 168 (6), July–August 2017, pp. 547–57.

Container gardening

See the Royal Horticultural Society(RHS) website for the many ways to grow edible plants in containers:

www.rhs.org.uk/advice/grow-your-own/containers/june-edible-container-idea.

www.rhs.org.uk/advice/grow-your-own/containers/veg-on-walls.

www.rhs.org.uk/advice/profile?PID=527.

Edible flowers

Bissell, Frances, *The Scented Kitchen: Cooking with Flowers*, New York, Serif, 2007.

Sullivan, Rebecca, *The Art of Edible Flowers: Recipes and Ideas for Floral Salads, Drinks, Desserts and More*, London, Kyle Books, 2018.

Mood-enhancing greenery

Lee, A., Jordan, H. C., and Horsley, J., 'The value of urban green spaces in promoting healthy living and well-being: prospects for planning', *Risk Management and Healthcare Policy*, Vol. 8, August 2015, pp. 131–7.

Maller, C., et al., 'Healthy nature healthy people: "contact with nature" as an upstream health promotion intervention for populations', *Health Promotion International*, Vol. 21(1), pp. 45–54.

Ulrich, R. S., 'Effects of gardens on health outcomes: theory and research', in Cooper-Marcus, C., and Barnes, M. (eds.), *Healing Gardens: Therapeutic Benefits and Design Recommendations*, New York, John Wiley, 1999.

Walking

Bratman, G. N., et al., 'Nature experience reduces rumination and subgenual prefrontal cortex activation', *PNAS*, Vol. 112 (28), July 2015, pp. 8567–72.

Hanson, S., and Jones, A., 'Is there evidence that walking groups have health benefits? A systematic review and meta-analysis', *British Journal of Sports Medicine*, Vol. 49 (11), 2015, pp. 710–15.

*

C3 Collaborating for Health, The Benefits of Regular Walking for Health: www.c3health.org/wp-content/uploads/2017/07/C3-report-on-walking-v-1-20120911.pdf (September 2015).

Wild swimming

Bongiorno, P., 'Cold splash – hydrotherapy for depression and anxiety', *Psychology Today*, 6 July 2014.

Mooventhan, A., and Nivethitha, L., 'Scientific evidence-based effects of hydrotherapy on various systems of the body', *North American Journal of Medicinal Sciences*, Vol. 6 (5), May 2014, pp. 199–209.

Siems, W. G., et al., 'Uric acid and glutathione levels during short-term whole body cold exposure', *Free Radical Biology and Medicine*, Vol. 16 (3), March 1994, pp. 299–305.

Mazes

Eliot, H., *Follow This Thread*, London, Particular Books, 2018.

*

The Maze Project: www.themazeproject.co.uk – a selection of unique mazes to visit in the UK.

Storytelling events and festivals

Beyond the Border in Wales – June: www.beyondthe border.com/festival.
Oxford Storytelling Festival – August: www.oxfordstory tellingfestival.co.uk.
The Society for Storytelling: https://sfs.org.uk – supporting and promoting storytelling in the UK.
Spark: http://stories.co.uk – connecting people through true stories.

Summer solstice

Vargas, Patti, and Gulling, Rich, *Making Wild Wines and Meads*, USA, North Adams, Storey Books, 2000.

*

Mead recipe from Robin Harford, wild food researcher and educator: www.eatweeds.co.uk.

Friendship Days

Chopik, W. J., 'Associations among relational values, support, health, and well-being across the adult lifespan', *Personal Relationships*, Vol. 24, April 2017, pp. 408–22.

Holt-Lunstad, Julianne, et al., 'Advancing social connection as a public health priority in the United States', *American Psychologist*, Special Issue, September 2017.

Lammas or Lughnasadh

Real Bread Campaign: www.sustainweb.org/realbread – helping you to seek out places to buy all-natural loaves locally.

The Sourdough School: www.sourdough.co.uk – sourdough tips, recipes, techniques and courses.

Autumn

Walker, M., *Why We Sleep: The New Science of Sleep and Dreams*, London, Allen Lane, 2018.

Mushrooms

Jayachandran, M., et al., 'A critical review on health promoting benefits of edible mushrooms through gut microbiota', *International Journal of Molecular Sciences*, Vol. 18 (9), September 2017, p. 1934.

Phillips, Roger, *Mushrooms*, London, Macmillan, 2006.

*

Wild Food UK: www.wildfooduk.com/articles/wild-mushrooms-for-beginners – safest wild mushrooms for the novice forager.

Mushroom walks and workshops with Andy Overall: www.fungitobewith.org.

Pulses

Kouris-Blazos, Antigone, and Belski, Regina, 'Health benefits of legumes and pulses with a focus on Australian sweet lupins', *Asia Pacific Journal of Clinical Nutrition*, Vol. 25 (1), 2016.

*

Health benefits of pulses: www.fao.org/3/a-i5388e.pdf.

The Pluses of Pulses: www.berkeleywellness.com/healthy-eating/food/article/pluses-pulses.

Challenges of the season

Metz, Jane A., 'Influenza and humidity: why a bit more damp may be good for you!', *Journal of Infection*, Vol. 71 (1), April 2015, pp. 54–8.

Pyankov, Oleg, 'Inactivation of airborne influenza virus by tea tree and eucalyptus oils', *Aerosol Science and Technology*, Vol. 46 (12), June 2012, pp. 1295–1302.

*

Condair: www.condair.co.uk/humidity-health-wellbeing/dry-air-and-airborne-infection – advice on humidity, air quality, health and well-being.

Healing flowers

Plantmaps: www.plantmaps.com/interactive-united-kingdom-plant-hardiness-zone-map-celsius.php/ and http://www.plantmaps.com/interactive-united-kingdom-first-frost-date-map.php – interactive maps showing dates of plant hardiness and first frosts.

Wildlife in the garden

BBC Earth: www.bbc.co.uk/nature – advice on supporting nature.

The British Trust for Ornithology: www.bto.org/about-birds/nnbw/make-a-nest-box – how to put up a bird box.

Ponds

Million Pond Project: freshwaterhabitats.org.uk/projects/million-ponds – Pond Conservation's Million Ponds

Project aims to create a network of ponds across the UK over the next 50 years.

Plants for a Future: pfaf.org/user/cmspage.aspx?pageid= 79/ – advice on creating an edible pond and bog garden.

Urban ponds and biodiversity, interview with Dr Ian Thornhill, University of Birmingham, 2012: www. birmingham.ac.uk/accessibility/transcripts/Ian-Thornhill-urban-ponds-biodiversity.

Well-being, movement and social connection

Common Ground: www.commonground.org.uk/ – imaginative ways to engage people with their local environment.

The natural world

Surfers against Sewage: www.sas.org.uk/our-work/beach-cleans/ – charity dedicated to the protection of oceans, waves, beaches and wildlife.

Understanding plastic packaging and the language we use to describe it: www.wrap.org.uk/sites/files/wrap/ Understanding%20plastic%20packaging%20FINAL. pdf.

Morning pages

Cameron, Julia, *The Artist's Way*, London, Macmillan, new edition, 2016.

Apple Day

www.commonground.org.uk/apple-day.

Winter

Embrace the cold

Buijze, Geert A., et al., 'The effect of cold showering on health and work: a randomized controlled trial', *PLoS ONE*, September 2016.

Castellani, John W., and Young, Andrew J., 'Human physiological responses to cold exposure: Acute responses and acclimatization to prolonged exposure', *Autonomic Neuroscience*, Vol. 196 (16), April 2016, pp. 63–75.

Keil, G., et al., 'Being cool: how body temperature influences ageing and longevity', *Biogerontology*, Vol. 16 (4), August 2015, pp. 383–97.

Knapton, Sarah, 'Open your bedroom window at night to prevent obesity and type 2 diabetes, says Oxford prof', March 2017: www.telegraph.co.uk/science/2017/03/20/open-bedroom-window-night-prevent-obesity-type-2-diabetes-says/.

Shevchuk, N. A., 'Adapted cold shower as a potential treatment for depression', *Medical Hypotheses*, Vol. 70 (5), 2008, pp. 995–1001.

Winter sun and snow

World Health Organization (WHO): www.who.int/uv/faq/whatisuv/en/index3.html – which environmental factors affect a person's UV exposure.

Viruses

Ben-Arye, Eran, et al., 'Treatment of upper respiratory tract infections in primary care: a randomized study using aromatic herbs', *Evidence-Based Complementary and Alternative Medicine*, 2011; online 2010, Nov 1. doi: 10.1155/2011/690346 PMID: 21052500.

Ryu, S., et al., 'Inactivation of norovirus and surrogates by natural phytochemicals and bioactive substances', *Molecular Nutrition and Food Research*, Vol. 59 (1), January 2015, pp. 65–74.

Circadian rhythms

Horne, J. A., and Reid, A. J., 'Night-time sleep EEG changes following body heating in a warm bath', *Electroencephalography and Clinical Neurophysiology*, Vol. 60 (2), February 1985, pp. 154–7.

*

'4-7-8 breathing' by Dr Andrew Weil: www.drweil.com/
    health-wellness/body-mind-spirit/stress-anxiety/
    breathing-three-exercises/2 – three breathing exer-
    cises and techniques.

Microgreens

Larkcom, Joy, *Grow Your Own Vegetables*, London, Frances
    Lincoln, 2002.
Larkcom, Joy, *The Organic Salad Garden*, London, Frances
    Lincoln, 2003.

Composting

Wiggly Wigglers: www.wigglywigglers.co.uk/composting/
    bokashi.html – tips and resources for bokashi com-
    posting.

Purifying your home

Ferris, Robert, 'Indoor air can be deadlier than outdoor
    air, research shows', *CNBC*, 22 April 2016.

*

Royal Horticultural Society: www.rhs.org.uk/advice/profile?
    PID=949 – houseplants to support human health.

Christmas trees

The Forest Stewardship Council: www.fsc-uk.org/en-uk – promoting responsible management of the world's forests.

The Soil Association: www.soilassociation.org/ – championing organic principles and practice.

Farmers' markets

Campaign to Protect Rural England: www.cpre.org.uk/resources/farming-and-food/local-foods/item/2897-from-field-to-fork – report on the importance and benefits of local food webs and farmers' markets.

Farma: www.farma.org.uk/members-map/ – non-profit association of the best real farm shops and real farmers' markets across the UK.

London Farmers' Markets: www.lfm.org.uk – resource for finding farmers' markets in London.

Reading

Elderkin, Susan, *The Novel Cure: An A to Z of Literary Remedies*, Edinburgh, Canongate, 2013.

Sieghart, William, *The Poetry Pharmacy: Tried-and-True Prescriptions for the Heart, Mind and Soul*, London, Particular Books, 2017.

*

The Reader: www.thereader.org.uk – resource pioneering the use of Shared Reading to improve well-being, reduce social isolation and build resilience.

Reading Well: https://reading-well.org.uk/books/mood-boosting-books – promotes self-help reading and has a list of mood-boosting books.

Wildlife

Mytting, Lars, *Norwegian Wood: Chopping, Stacking and Drying Wood the Scandinavian Way*, London, MacLehose Press, 2015.

*

Wilderness or bushcraft organizations such as:

The Forest School Association: www.forestschoolassociation.org/what-is-forest-school.

Natural Pathways: www.natural-pathways.co.uk/footsteps-tracking.php – offers courses in tracking and other wildlife experiences.

The Woodland Trust: www.woodlandtrust.org.uk – wilderness organization.

Awe

Bai, Y., et al., 'Awe, the diminished self, and collective engagement: Universals and cultural variations in the small self', *Journal of Personality and Social Psychology*, Vol. 113 (2), August 2017, pp. 185–209.

Henderson, Caspar, *A New Map of Wonders: A Journey in Search of Modern Marvels*, London, Granta, 2018.

Stellar, Jennifer, et al., 'Self-transcendent emotions and their social functions: compassion, gratitude, and awe bind us to others through prosociality', *Emotion Review*, Vol. 9 (3), July 2017.

Weber, Andreas, *The Biology of Wonder: Aliveness, Feeling, and the Metamorphosis of Science*, Canada, New Society Publishers, 2016.

*

'The Benefits of Feeling Awe', 2016 interview with Craig Anderson, University of California, working with Dacher Keltner on Project Awe: https://greatergood.berkeley.edu/article/item/the_benefits_of_feeling_awe.

Thanks

Several people have provided invaluable support, first and foremost Tessa Strickland for her experience and wisdom. Special thanks, too, go to dear friends: Hilary Arnold and Annabel Huxley for encouraging me, along with Ruth Thomson, Piers Allen and Carolyn Steel and my fellow 'sibaritas'.

For inspiration and a deeper understanding of plants and nature's extraordinary bounty, I owe much to the late Frank Cook, and to Robin Harford (and for his mead recipe), Miles Irving and Ray Mears.

I have Simon Mills and his book *The Essential Book of Herbal Medicine* to thank for my spur towards professional herbal medicine, and the recognition that health is about ecology and community, touched by all aspects of life.

My thanks above all go to my editor, Aimée Longos, for finding me in the first place through Living Medicine and Chelsea Physic Garden. She was a delight to work with, as was the team of Nikki Sims, Emily Robertson, Annie Lee, Ellie Smith, Francisca Monteiro, Caroline Wilding, and Peach Doble with her beautiful illustrations.

Index